Technology and Psychological Well-being

In the modern world we are surrounded by technology. Gadgets such as cell phones, portable computers, and electronic diaries accompany us throughout the day. But is this a good thing? Are we being served by these technological wonders, or have we become enslaved by them? Does constant availability via technology make us more efficient or more stressed? Is our ability to connect with others all over the world, day or night, making us more sociable or turning us into recluses in a virtual world? This book considers the impact of technology on the different spheres of our life – work, home, family, and leisure – and assesses ways in which to build better communication between technology developers and society to ensure that technology enhances our lives and psychological well-being, rather than damaging them.

YAIR AMICHAI-HAMBURGER is the Director of the Research Center for Internet Psychology at the Sammy Ofer School of Communications at the Interdisciplinary Center (IDC), Herzliya, Israel.

Technology and Psychological Well-being

Edited by
Yair Amichai-Hamburger

CAMBRIDGE
UNIVERSITY PRESS

CAMBRIDGE UNIVERSITY PRESS
Cambridge, New York, Melbourne, Madrid, Cape Town, Singapore,
São Paulo, Delhi

Cambridge University Press
The Edinburgh Building, Cambridge CB2 8RU, UK

Published in the United States of America by Cambridge University Press,
New York

www.cambridge.org
Information on this title: www.cambridge.org/9780521885812

First published 2009

Printed in the United Kingdom at the University Press, Cambridge

A catalogue record for this publication is available from the British Library

Library of Congress Cataloguing in Publication data
Technology and psychological well-being / edited by Yair Amichai-Hamburger.
p. cm.
Includes index.
ISBN 978-0-521-88581-2 (hbk.)
1. Technological innovations – Social aspects. 2. Technological
innovations – Psychological aspects. 3. Technology – Social aspects.
4. Technology – Psychological aspects. I. Amichai-Hamburger, Yair.
II. Title.
HM846.T436 2009
303.48′3 – dc22 2009016623

ISBN 978-0-521-88581-2 hardback

Dedicated to Debbie – my wife, my partner,
my friend
And for our wonderful children – Micha’el, Talia,
Keren, and Yaron

Contents

Figures

Tables

Contributors

YAIR AMICHAI-HAMBURGER: The Research Center for Internet Psychology (CIP), Sammy Ofer School of Communications, The Interdisciplinary Center (IDC), Israel

AZY BARAK: Department of Counseling & Human Development, University of Haifa, Israel

CAROLINE BIRON: Lancaster University Management School, Lancaster University, UK

RAE LESSER BLUMBERG: Sociology Department, University of Virginia, USA

CARY L. COOPER: Lancaster University Management School, Lancaster University, UK

SHERRY COULSON: Department of Physical Medicine and Rehabilitation, The University of Western Ontario, Canada

SUSAN C. EATON: Kennedy School of Government, Harvard University, USA

KATHERINE G. FRY: Department of Television and Radio, Brooklyn College of CUNY, USA

STEVEN L. GOLDMAN: Departments of Philosophy and History, Lehigh University, USA

OSCAR I. GONZALEZ: Department of Psychology and Social Behavior, University of California, Irvine, USA

PETER A. HANCOCK: Department of Psychology, University of Central Florida, USA

JEFFREY W. JUTAI: Faculty of Health Sciences, University of Ottawa, Canada

ELLEN ERNST KOSSEK: School of Labor and Industrial Relations, Michigan State University, USA

BRENDA A. LAUTSCH: Faculty of Business Administration, Simon Fraser University, Canada

RAYMOND W. NOVACO: School of Social Ecology, University of California, Irvine, USA

MICHAEL P. O'DRISCOLL: Department of Psychology, University of Waikato, New Zealand

TAL ORON-GILAD: Department of Industrial Engineering and Management, Ben-Gurion University of the Negev, Israel

GEORGE RODMAN: Department of Television and Radio, Brooklyn College of CUNY, USA

ELIZABETH RUSSELL-MINDA: Department of Physical Medicine and Rehabilitation, The University of Western Ontario, Canada

Acknowledgements

Without the help of a number of people this book could not have come into being. My grateful thanks to the authors for their hard work and excellent contributions. My gratitude to those whose comments and encouragement were invaluable: Albert Auster, Susan B. Barnes, Kent Campbell, Joanne Cohoon, Randall Collins, Lori Foster Thompson, Shaul Fox, Efrat Gil, David Haken, Tsahi Hayat, Matthew Kalman, Tal Katz-Navon, Meni Koslowski, Moshe Krausz, Ann Lane, Gerhard E. Lenski, Ingvor Pettersson, Rivka Ribak, Jason T. Siegel, Louise Sylvester, Keith Weber, Patrice L. (Tamar) Weiss, and Mina Westman.

My thanks go to Andrew Peart and Carrie Cheek at Cambridge University Press, who helped me materialize my ideas into a book. And a special thank-you to my beloved wife Debbie, without whose ideas and support this book would never have seen the light of day.

Introduction

Today, it is hard to find a corner in our lives that is not affected by technology. We are surrounded by it at home, commuting, at work, and in our leisure time. Given its dominant position, it should come as no surprise that technology has a strong impact on our well-being. In recent years, there have been some major studies into different aspects of technology and its influence on our well-being. However, until now, no book has attempted a wide-ranging appraisal of how the technology that is such a part of our everyday lives impacts our well-being. This is precisely the aim of this book.

The term "technology" has its origins in the Greek concept *technologia*, which includes the ideas *techne* – "craft" – and *logia* – "saying." Many people have attempted to give this concept a definition. For example, Stein (1966) defined it as the sum of the ways in which a social group provides itself with the material objects of its civilization. Rousseau (1978) suggested that technology is the process used to transform raw materials into end products. Kipnis (1991) defined it as the use of systematic procedures to produce intended effects. However, a definitive meaning of the term "technology" is elusive: the term may be used to refer to material objects of use to humanity, such as hardware, machines, or utensils, but it can also encompass broader components, including methods of organization or techniques. The term can either be applied generically or to specific areas. Generally speaking, technology is the relationship that society has with its tools and crafts.

Well-being is also a term that defies a single definition. It has been used interchangeably with such concepts as happiness, health, welfare, comfort, security, and safety. In addition, there are associated terms, for example psychological well-being, subjective well-being, and so on. Argyle (1992) suggested that when people are asked to define happiness, they answer in one of two ways: Some describe happiness in terms of a positive emotion, like joy, while others will describe it in terms of contentment and satisfaction with life. Helliwell and Putnam (2004) distinguish between happiness and life satisfaction. They argue that: "Generally speaking, self-ratings of 'happiness' turn out to reflect relatively short-term, situation-dependent (affective) expressions of mood, whereas self-ratings of 'life satisfaction' appear to measure longer-term, more stable (cognitive) evaluations" (2004: 1435). Conversely, Ben-Shahar (2007) believes both the emotional component and the life-satisfaction component should be within the definition; thus, he defines happiness as the "overall experience of pleasure and meaning." In other words, a happy person enjoys positive emotions while perceiving his or her life to be purposeful. This definition does not pertain to a single moment, but rather to a generalized aggregate of one's experiences: A person can endure pain at times and still be happy overall.

Among other terms used to describe well-being is "quality of life." Janse *et al.* (2004) suggested that quality of life is a multidimensional construct; included within it are physical, emotional, mental, social, and behavioral components. Schwarz and Strack (1999) argued that subjective well-being can be defined as the individual's current evaluation of his or her happiness. Such an evaluation is often expressed in affective terms: When asked about subjective well-being, participants will often reply, "I feel good" (Schwarz and Strack, 1999). Pollard and Lee (2003) believe that well-being is such a complex construct that, despite all their attempts, researchers have never managed to find accurate ways to delineate it. The variation among the definitions employed, even within an individual discipline, is so great that producing a comprehensive overview of definitions in use within the literature is a formidable task. Nevertheless, it seems clear that what most definitions have in common is a concern for the psychological health of human beings.

In general, people want to feel that their happiness is justified; they actually appear to prefer the real feeling of enjoyment after they have worked hard to achieve it. This is a departure from the more hedonistic belief that enjoyment per se is the ultimate goal of individuals (Tatarkiewicz, 1976). In addition, it was found that people are ready to sacrifice short-lived happiness if it conflicts with the pursuance of other goals they value (Kim-Prieto, 2001).

Why is it important to study well-being? The benefits of well-being at the individual level are very significant. They are summed up by Diener and Seligman (2004) as (1) society: Democratic governance may be enhanced through the well-being of the populace; (2) income: Happy people earn higher incomes than unhappy individuals; (3) work: Happy and satisfied workers are more likely to perform higher citizenship organizational behavior (that is, individual effort that benefits the organization, unrecognized by the formal reward system). Satisfaction level in work units is correlated with high productivity and profitability; (4) physical health: Longevity is correlated with high levels of well-being. Individuals with low levels of well-being have compromised immune systems and are more likely to have certain diseases as compared with individuals with high levels of well-being; (5) mental disorders: Happy individuals score lower in mental disorders, as opposed to unhappy individuals; (6) social relationships: Positive well-being is associated with a higher likelihood of getting married and staying happily married, and with having large numbers of friends and strong social support.

Lyubomirsky, King, and Diener (2005) found that happy people are more sociable, generous, creative, active, tolerant, healthy, altruistic, economically productive, and long-lived. It seems that the promotion of well-being is not just an important end itself, but also leads to positive outcomes for society.

Since well-being is so important to our society and since technology seems to pervade almost all areas of our lives, the study of the different effects of technology on our well-being would seem fundamental. In this book we have concentrated on major subject areas, including communication, Internet, medicine, transportation, gender, human factors, work environment, information society, and the work–home balance.

In the first chapter, George Rodman and Katherine G. Fry examine communication technologies and well-being. They suggest that communication technology creates a Yin and Yang in the sense that new devices can have positive or detrimental effects on psychological well-being. They conclude that, in terms of psychological impact, there appears to be a golden mean pertaining to the use of technology after which, the psychological benefits having been maximized at this midpoint, a downward spiral of decline begins.

The authors examine the relationship between communication technology and psychological well-being from a twofold perspective. They look at the framework of history and particularly how changing communication technologies have fostered changes in the concept of well-being through four communication epochs: primary orality, writing, typography, and electronic communication. Next, they look at the framework of social-science media research and analysis, especially in terms of such perspectives as cultivation theory, media addiction, parasocial relationships, environmental mastery, uses and gratifications, and desensitization. They argue that it is important to study the concept of well-being and the dominant forms of communication within various eras, and suggest that further research into this question, from both historical and social-scientific perspectives, would be valuable.

In the second chapter, Yair Amichai-Hamburger and Azy Barak discuss the impact of the Internet on our well-being. They describe the unique components of the Internet environment, such as greater anonymity, the diminution of the importance of physical appearance, greater control over the time, place, and pace of interactions, and the ease of finding similar others and their psychological impact on users. They focus on the influence of the Internet on our different life spheres. On the personal level, they assess the role of the Internet as a vehicle for revealing the "real me" and the "true self" – that is, as a means of reframing individual and subjective identities. The authors go on to examine the effects of the Internet environment on different areas of life. On the interpersonal level, they include discussions of the patient–physician relationship, online psychotherapy and counseling, and online support groups. On the group level, they focus on stigmatized groups, support groups versus treatment groups, online contact between rival groups, religious groups, and solving the conflict between belonging to a group and individuality. They conclude with some suggestions for further research to promote human well-being.

In Chapter 3, Steven L. Goldman examines the topic of information, innovation, and society. Assessing the impact on human well-being of recent innovations in digital information technologies requires clarifying the historical role played by information in Western societies, the meaning of the term "information," and the generic process by which technologies affect social change. In fact, there are at least three distinct usages of "information" extant today: as content, as a content-independent mathematical pattern, and as an elementary feature of physical reality. Contrary to the popular notion of an information revolution currently transforming societies around the world, only the second, and least familiar, of these usages merits being called "revolutionary." The media-reinforced popular notion obscures the evolutionary character of the role of information as content in society and the identification of information as a causally efficacious feature of the world. The claim that technological innovations unilaterally drive social change obscures the dialectical character of the technology–society relationship and the criteria for judging which social changes constitute improvements.

Over the last two decades, there has been a global explosion of technology, particularly computer-based and other media-based technologies. The utilization of information and communication technologies has become so pervasive that people can no longer conceive of what work and life would be like without them. In the workplace, the vast and rapid introduction of these technologies has brought many changes and widely influenced the organization of work, the psychosocial work environment, and work practices.

In the fourth chapter, Michael P. O'Driscoll, Caroline Biron, and Cary L. Cooper examine the topic of technological change and psychological well-being. The authors discuss some of the major psychosocial impacts which this utilization of information and communication technology (ICT) has wrought on the psychological well-being of workers. One positive impact of ICT is increased flexibility by reducing the need for traditional offices. This can have positive benefits for individuals endeavoring to balance work and family responsibilities, and for organizations in terms of adapting more effectively to constant changes in the global economy. However, many studies have shown that the increased utilization of ICT can have both a positive and a detrimental impact on workers' well-being. Research has illustrated that technological changes are closely related to job demands, job control, and social relationships in the workplace, which in turn are critical factors for employees' stress and well-being.

When technologies are introduced in the work environment, adequate thought has to be given to creating mechanisms that ensure that workers participate in the decision-making process related to the introduction and utilization of ICT, and that perceptions of control and mastery over the technologies are enhanced. The design, development, and implementation of ICT require a thorough understanding of how people think, feel, and respond when dealing with systems that are challenging to cope with and which create significant changes in the work environment.

In the fifth chapter, Tal Oron-Gilad and Peter A. Hancock assess the development from ergonomics to hedonomics – trends in human factors and technology. Their chapter examines the transition in focus within the human factors engineering domain (Human Factors Engineering is the discipline of applying what is known about human capabilities and limitations to the design of products, processes, systems, and work environments) from the prevention of pain (ergonomics) to the promotion of pleasure and using technology to facilitate well-being (hedonomics). This transition is coupled with technological advancements and their penetration into a wide range of user groups. Implications for workers' well-being are discussed. Difficulties in implementing hedonic principles are outlined, along with examples of emerging areas of research.

Ellen Ernst Kossek, Brenda A. Lautsch, and Susan C. Eaton consider the conditions necessary for teleworking to enhance employees' well-being. In Chapter 6, they look at professionals' use of telecommuting as a lens through which to examine the ways in which access and the psychological experience of using telework relate to types of job flexibility control, boundary management, and work and family effectiveness. The results of survey and interview data from a sample of 316 professionals, some of whom teleworked and some of whom did not, show that mere access to teleworking will not necessarily lead to positive work and family outcomes, and sometimes may have null or even negative effects. The authors consider that these issues need to be more widely discussed in the work and family and human resource management literatures. It is important to note that for the professionals they studied, job flexibility control has a much stronger favorable relationship to turnover intentions than it has to work–family balance. It may be that telework access is insufficient to lower work–family conflict for professional jobs. Job control and boundary management are important moderators of well-being.

In the seventh chapter, Raymond W. Novaco and Oscar I. Gonzalez tackle the topic of commuting and well-being. Commuting stress, whether associated with driving on congested roads or with strains in public transport, has great relevance for the health and social sciences, urban planning, engineering, economics, and business management. The authors provide an overview of the research on commuting stress and the latter's impact on physical health and psychological adjustment. Within this framework, they assess historical developments in automobile commuting and technologically driven alternatives to cope with commuting stress related to personal health, work performance, and family relationships.

The chapter continues with a discussion on the subsidiary subject of driver stress within the context of its association with violations of the law and accidents. Comparative studies across commute modes and information-age innovations are also discussed. The authors cite evidence in support of telecommuting having become a coping strategy as the convergence of technological, economic, social, and psychological factors have made work less place-dependent. They review the major variables that moderate commuting stress, such as personal control, commute predictability, and gender, as well as environmental factors affecting commuting stress and coping. The authors conclude that despite the adversities of commuting, some of its elements can serve as enhancements to well-being, such as privacy, protected time, and the symbolic value of personal vehicles and freedom. They conclude with a discussion of adaptation as an overarching concept.

In the chapter that follows, Jeffrey W. Jutai, Sherry Coulson, and Elizabeth Russell-Minda deal with technology and medicine. Medical technology in the broadest sense refers to the diagnostic or therapeutic application of science and technology to improve the management of health conditions, and describes a vast array of devices and techniques. It is generally assumed that virtually every form of medical technology has a positive impact on quality of life and well-being, but it is important to analyze this assumption. This chapter examines how the relationship between technology and health is conceptualized and, therefore, how effectively it might be researched. It then reviews the research evidence for the impact on quality of life and well-being of technologies to assist mobility and vision for individuals with disabilities. It concludes with recommendations on how assistive technologies might be better researched and developed to enhance well-being.

In the penultimate chapter, Rae Lesser Blumberg examines the issue of gender and technology from a feminist, anthropological perspective. She argues that, throughout human history, there has been a greater contribution by women to subsistence technology and ensuing societal wealth and well-being than is commonly recognized. From baby-carrying slings and gathering baskets among our foraging ancestors to the development of cultivation, to the little-heralded near-gender-parity in the life sciences that are driving much of today's technological progress, women's roles have been more important than renowned. Recent gender and technology/science studies have focused more on the computer end of information technology, where the proportion of undergraduate women has been dropping. This reinforces the inaccurate view that women play a secondary role in technology and leads to the possible detriment to their well-being. The author aims to tell a fuller story, including women's roles in the emerging "techno-economic base of knowledge and life."

In the last chapter, Yair Amichai-Hamburger adopts a wide-ranging perspective on the issue of technology and well-being. The chapter begins by assessing the major roles played by technology in our lives. Is technology a means to an end or does it serve a function in and of itself? Does technology carry within it the answers to many of the current and future challenges facing society or is it a tool that has to be directed in order to achieve our real goals? This chapter initially explores some of the more worrying aspects of our highly technological society and then goes on to discuss ways in which present technological achievements and those of the future may be harnessed to promote the well-being of the societies in which they operate.

This book aims to create a more comprehensive understanding of the impact of technology on well-being, to place this issue firmly on the agenda, and to provide tools to improve the positive impact of technology on well-being. We believe that with the creation of the "global village" and the technological explosion, the issue of how technology impacts our well-being has become crucial throughout the world. We hope that this book will play a part in raising awareness of this critical universal concern.

References

Argyle, M. (1992). *The Social Psychology of Everyday Life*. New York: Routledge.

Ben-Shahar, T. (2007). *Happier*. New York: McGraw-Hill.

Diener, E., and Seligman, M. E. P. (2004). Beyond money: Toward an economy of well-being. *Psychological Science in the Public Interest*, 5, 1–31.

Helliwell, J., and Putnam, R. D. (2004). *The Social Context of Well-being*, Series B, 359 (pp. 1435–1446). London: Philosophical Transactions of the Royal Society.

Janse, A. J., Gemke, R. J., Uiterwaal, C. S., Van Der Tweel, I., Kimpen, J. L., and Sinnema, G. (2004). Quality of life: Patients and doctors don't always agree: A meta-analysis. *Journal of Clinical Epidemiology*, 57, 653–661.

Kim-Prieto, C. Y. (2001). What's a wonderful life? The pursuit of personal happiness versus the pursuit of in-group obligations. Unpublished Master's thesis, University of Illinois at Urbana-Champaign.

Kipnis, D. (1991). The technological perspective. *Psychological Science*, 2, 62–69.

Lyubomirsky, S., King, L. A., and Diener, E. (2005). The benefits of frequent positive affect: Does happiness lead to success? *Psychological Bulletin*, 131, 803–855.

Pollard, E., and Lee, P. D. (2003). Child well-being: A systematic review of the literature. *Social Indicators Research*, 61, 59–78.

Rousseau, D. M. (1978). Measures of technology as predictors of attitudes. *Journal of Applied Psychology*, 63, 213–218.

Schwarz, N., and Strack, F. (1999). Reports of subjective well-being: Judgmental processes and their methodological implications. In D. Kahneman, E. Diener, and N. Schwarz (Eds.), *Well-being: The foundations of hedonic psychology* (pp. 61–84). New York: Russell-Sage.

Stein, J. (Ed.) (1966). *Random House Dictionary of the English Language*. New York: Random House.

Tatarkiewicz, W. (1976). *Analysis of Happiness*. The Hague, Netherlands: Martinus Nijhoff.

1 Communication technology and psychological well-being: Yin, Yang, and the golden mean of media effects

George Rodman and Katherine G. Fry

The Yin and Yang of ancient Chinese philosophy is the symbol of two primal opposing but complementary forces found in all things in the universe. The Yin and Yang of communication technology could be described as follows: like most technological advances, communication devices improve people's lives immeasurably. Once they use them, people cannot imagine living without them. On the other hand, critics contend that these same electronic communication media – beginning with movies, television, and radio, and today including the Internet, video games, cell phones, PDAs (personal data assistants), and other mobile devices – are turning those of us who spend an inordinate amount of time with them into isolated, narcissistic, anxious, and ultimately unhappy beings. Whether we are technically addicted, or just spend too much of our time communing with digital visual, audio, and text devices, we are thought to be socially fragmented, less able than others who are not as drawn to these technologies to connect in face-to-face interactions. We experience, perhaps, a diminished sense of psychological well-being in comparison to them.

It seems safe to say that each new communication technology could be either a detriment or a benefit to psychological well-being. The Yang of detriments and the Yin of benefits exist as a whole. We could add Aristotle's conception of the golden mean to this equation: There is a midpoint between extremes in which the psychological benefits of communication media are maximized. Below this point, one does not experience the benefits. Above this point, one experiences the detriments.

To cite one example, the Yin and Yang of media effects are seen in media's impact on self-acceptance, the quality of having a positive attitude toward oneself and one's life. Communication media such as home movies and home videos have been used to record our lives as a series of golden moments and celebrations, presumably having a positive effect. However, children and teens compare themselves to idealized media images, which has been blamed for eating disorders (David and Johnson, 1998).

Another example: The Yin and Yang of media technology can be seen in the way portable music players such as the iPod are used. Their huge popularity testifies to the great pleasure they provide, but they also isolate listeners by allowing them to tune out the world around them. Parents today notice that their children no longer squabble about what station the car radio should be tuned to. Each family member is tethered to his or her individual earplugs, and the eerie silence has reminded more than one dad of an invasion of pod-people. The absence of squabbling is comforting, but one cannot help feel that opportunities for bonding are being lost with each mile.

In this chapter we will take a look at the Yin, the Yang, and the golden mean of communication technology both from a historical and a social-science point of view.

Approaches to communication technology and psychological well-being

Psychological well-being is a theoretical concept that refers to an individual state of psychological equilibrium. Also known as mental health, subjective well-being, or self-reported happiness, psychological well-being has been the subject of scientific study for more than a century (Brim, 1994). Until recently, most studies of well-being defined mental health as not being sick, as an absence of anxiety, depression, or other forms of psychological problems. The modern conception of psychological well-being emphasizes positive characteristics of growth and development. Ryff (1995) points out that this concept involves six commonly accepted components: self-acceptance, purpose in life, environmental mastery, personal growth, positive relations with others, and autonomy. Amichai-Hamburger (2005) explains the state of well-being by reference to pioneers Jung and Rogers. Drawing on Jung's theory (1939), psychological well-being is understood as the successful creation of a balance between the opposing forces of introversion and extroversion in one's personality. Carl Rogers (1980) theorized that there are various forms of self – the true self, the ideal self, and the self-concept. Balance, he argued, is achieved when there exists the smallest gap between these three states. One could argue, then, that a state of psychological well-being is the state wherein there exists such an internal balance. It seems these descriptions of well-being define a state of equilibrium. Such equilibrium is determined as a measure of social connectedness. Therefore, social connections are an important component for the balanced individual who has achieved psychological well-being.

Are our electronic media today eroding our sense of well-being by throwing off that balance? Do children and adults spend too much time playing video games, immersing themselves in different virtual worlds, or talking on the cell phone? Is it electronic media, specifically, that have us concerned? Have parents, scholars, or other interested parties ever been concerned with young people, or anyone, really, who spends too much time reading or writing to the detriment of their interpersonal, face-to-face interactions? Early church leaders were concerned that printed bibles would unsettle the masses and confuse them about matters of church doctrine. Even today there are parents who worry that their children are becoming unpopular bookworms. Generally, however, we tend to value literacy and encourage reading and writing over most other activities. But the kind of literacy needed to navigate the plethora of electronic communication technologies, and the immersion required in an environment of multisensory modes – video, sound, and text – are not so highly prized. In fact, they seem to leave many of us concerned about possible negative, psychologically unhealthy side effects.

It may be, however, that the balance is changing. In other words, it may be that as our communication technologies change, we change with them. Currently we are experiencing, because of our changing media environment, a seismic shift in the economy, in politics, and in social transactions of all sorts. We do not know, yet, what all of the psychological implications will be, but we have historic precedent that shows us that we change how we understand the world and ourselves when the means of communication change. Beginning with the telephone and the telegraph in the nineteenth century, electronic communication has continuously developed and offered us new ways to interact with more people at a time than was possible for centuries before that (Innis, 1950). Besides collapsing space and time, and thus the way people understood these concepts, the early electronic communication devices began a revolution that has had enormous social and psychological implications. The shift from a print-based society to an electronic media society did not happen overnight, and long before that particular shift in the dominant mode of communication there were others that also changed societies and individuals by changing the way social connections were made, thus how individuals interacted with and understood themselves and their relationship to the world.

There are different ways to approach the topic of communication technology and psychological well-being. Here we will look at the concept of psychological well-being from first the framework of history – Western history specifically – and particularly the history of changing communication technologies, and second, the framework of social-science media research and analysis. We are confining our discussion to these two frameworks because the theoretical concept of well-being has been developed within the social science of psychology. We chose to remain within that discursive framework while adding the historical approach to communication technology development because it adds, we believe, an important context to the contemporary social-science perspective. While critical cultural theories of media analysis in the traditions of the Frankfurt School and Birmingham School are valuable and could also add the other important contexts of political economy, the politics of representation, and audience reception, they are more culturally oriented (see Durham and Kellner, 2006; Horkheimer and Adorno, 2002). Therefore, in this initial foray into examining psychological well-being, we decided to begin by looking historically at the self as shaped by communication forms. Certainly within the social-science perspective well-being is an assumed positive state that one strives to achieve. However, within the historical framework, which emphasizes change over time, well-being is historically and culturally contingent. In the social-science approach media content and behavior/attitudes/opinions are examined in a cause–effect relationship. In the historical approach media forms or technologies are distinct from content, and are examined in terms of the social and cultural environments they engender by virtue of their different sensory biases.

In this chapter we first examine psychological well-being, placing the concept and definition in historical perspective by exploring the extent to which the current research into media and psychological well-being might be understood along a historical continuum of change. Well-being is not an immutable psychological fact. Rather, what it means to experience psychological well-being has changed over a very long period of time, corresponding to changes in our concepts of the self and others. The concepts of loneliness, anxiety, efficacy, and autonomy are relative to the eras within which they are named and understood. By using the historical perspective, we will examine how such notions might change over time, and how they are connected with changes in communication technologies.

Historical perspective

The social connections one experiences are formed through communication, which is, very generally, a process of symbolic exchange. Communication among humans takes many forms, and is achieved through many different symbolic means. These include spoken language, written language, physical gesture and other forms of body language, visuals, and sound. The process of communication in many instances takes place through forms of technology. Communication technologies can also be referred to as communication forms or communication media. Form and media will be used interchangeably here. The form in which a symbolic exchange takes place is a vital component in the communication itself. For example, speech is distinct from written language, not only because the former is biased toward the ear and the latter biased toward the eye, but also because the written word requires literacy, or knowledge of the written symbols, the grammar, and the syntax of a language, in order to be able to communicate meaningfully with others who share that same knowledge. Likewise, visuals communicate in a manner very different from that of words, shaping ideas in ways that are difficult to compare with language, and evoking very different responses (Stephens, 1998).

Social connections rely on communications that take place in many different contexts, from the face-to-face context of a personal conversation, to the electronically enhanced interpersonal context of a cell phone conversation, to the mass-mediated context of broadcast or cable television, or to the computer-mediated context of e-mail. If well-being is measured through social connections, forged through communications that are shaped by media, which are vital components of content and context, then one might wish to examine the ways in which different media operate uniquely in terms of social connections. As a result, returning to the level of the individual, one could examine how different media uniquely shape or interact with a sense of well-being. Note that there has been no reference to specific content so far. That is because, in this first section, the focus is on the media, the technologies, themselves as they have developed and been used essentially by Western societies. The media shape content. Indeed, one could argue they are more important than content, as media theorist Marshall McLuhan makes clear in his well-known aphorism *the medium is the message* (1964).

Historic changes in the concept of the individual

The notion of self has changed historically. Psychologist John Teske (2002) has traced the evolution of this concept. Teske is concerned about the contemporary paradox of expanding networks and interconnectedness but increasing sense of fragmentation. He acknowledges, and describes in detail, numerous studies in the emerging field of cyberpsychology, specifically those studies that examine interpersonal relationships and the Internet. Drawing on the work of Philip Cushman (1990) he also traces the history of individuality. Individuality, Teske argues, is a historically contingent concept, defined throughout time by philosophy and theology. Cushman himself traces historic shifts in understanding the self from the religious to the scientific. We have increasingly come to see the individual, Teske argues, as the locus of salvation. We value self-actualization, personal growth, and transformation. This is in contrast with earlier periods of time when community and interdependence were more highly valued and the sense of one's self as an individual was not considered as important. Teske compares the outward-looking Aeschylus, the Greek poet born in 525 BC, with the more tortured interiority of Saint Augustine, born in AD 354, and finally with the narcissistic F. Scott Fitzgerald of the 1920s US jazz era. Each of these figures is emblematic of individuals during his era, with an increasing shift in emphasis over time toward a sense of self as separate from others. The term "self," Teske points out, does not appear in the *Oxford English Dictionary* until 1595 (2002: 690). Teske describes a process by which individuals become increasingly secular and secluded. The result, he explains, is that today we nurture autonomy and self-sufficiency and see higher levels of anxiety concomitant with decreases in social connectedness, particularly within Western cultures, and more specifically within the United States.

Teske connects the increase in the sense of the importance of the individual with historic changes in the economy, from agriculture to industry, and more recently the postwar economy, as well as shifts from rural to urban living. Classic social theorists such as Durkheim (1964), Weber (1985), and Marx and Engels (1978) acknowledged an increasing sense of the individual as separate from society as each theorized, in quite different ways, a growing sense of personal alienation as Western industrial society became increasingly large and bureaucratic. During almost the same time period, Freud (1901) introduced Western culture to the ideas of various states of individual consciousness as well as the unconscious, further focusing our attention on the importance of the individual psyche. Since the late nineteenth century all of these theories have become mainstreamed and have had a profound effect on the way we have, subsequently, thought about individuals. All of these shifts in thinking about self are directly related to historic changes in communication technology.

Historical development of communication technologies

Scholars interested in media forms have examined the evolution of societies and cultures with regard to the dominant medium of a given era. Each historical era, or communication epoch, is characterized by the major means of communication within that epoch, and the psychological, social, economic, political, cultural, and epistemological environment emanating from the dominating communication form (Lum, 2006). The term environment here has clear ecological connotations. Each medium creates an environment in which communication takes place. In other words, each form of communication, because of its sensory bias (verbal, aural, visual, etc.), encourages a particular way of transmitting and accepting messages. In addition, each encourages a particular way of understanding the world, of organizing thought, and of developing relationships with others. For example, print has a visual bias, while speech has an aural bias. Print transmits messages and meaning through words on a page, while speech transmits messages through voice, including spoken words and the various intonations and verbalizations included in speaking. The distinguishing characteristics of each form of communication are important means by which to compare the vastly different ways in which each is used as a communication medium, creating vastly different environments when each is the dominant form of communication within a historic era. Our media environments inhabit us, and we can examine and compare the different social and psychological dimensions associated with media environments in different historical time periods.

First communication epoch – primary orality Scholars Walter Ong (1982) and Eric Havelock (1963) were interested in the profound differences between cultures that communicated primarily through oral means, and cultures that used writing systems. Havelock determined that linguistic verbalization within oral cultures was a kind of technology, and could be compared with the communication technologies that followed in subsequent communication epochs, such as writing, print, and then electronic media. For example, he examined Greek culture, looking at ancient Greek oral culture and comparing it to the culture and social organization once an alphabet and writing were introduced (Havelock, 1963: 4–43). While Ong also connected major shifts in a range of social and cultural phenomena to the shifting from one communication epoch to another, he was also interested in what he referred to as the psychodynamics engendered within communication epochs, particularly in the psychodynamics of oral cultures in contrast with the psychodynamics of literate cultures, or cultures that used writing systems. Drawing on the work of linguistic anthropologists, Ong asked, what are the psychological characteristics of people within primary oral cultures? How do such people think? How does this change the way people in oral cultures understand the world and themselves? He concluded that the oral mind possessed a very specific cognitive and emotional dimension. Briefly, he argued that individuals in oral cultures, because they relied on verbal communication, tended to organize information in their minds in ways that helped them to remember, since memory was vitally important as there was no way in which to record information. Ong's terms to describe the oral mind, such as additive, aggregate, and redundant, are all terms that describe the ways that information was taken in, understood, and stored (1982: 37–57).

In these nonalphabet societies, the world was conceived of as a whole, composed of meaningful interrelations, fluid interactions, and cycles, as opposed to discrete, abstract, and hierarchical ideas and pieces of information – the artifacts of literacy. Face-to-face communication and the immediate social and natural environment are of primary importance in an oral culture. In such a specific social organization, a sense of well-being would no doubt be linked more to the cohesiveness of the social collective than to separate individuals. Privacy would not be understood as an individual value, if it would exist at all. The sense of oneself as an individual did not resonate with primarily oral people. Personal autonomy in a psychological sense did not exist. Teske (2002) explains that notions of privacy and individuality did not become important until after about AD 1500. In an oral society well-being would be determined by the social body in terms of complete interconnectedness. A separation between the social and the individual begins with the inventions of alphabets and writing.

Second communication epoch – writing The first alphabets were introduced between 900 and 700 BC, marking the beginnings of literate societies. Writing systems, print on paper via abstract alphabet systems, became a barrier to direct experience as opposed to oral societies where all communication happened through direct experience with the environment and with others face to face. Writing wrought enormous changes in the individual's relationship to society. Reading and writing encourage a sense of individualism because they introduce privacy – private thoughts, the privacy necessary to read and write, and a general sense of interiority. This sort of privacy, a focus on one's individual thoughts, allows one to develop a separate sense of self. Personal autonomy develops. Writing also encourages rational thought, and allows for written records, thus creating a history that can be stored and accessed in books and other written materials. Writing, or literacy, also more equitably disperses knowledge throughout a society, displacing the bearers of knowledge and wisdom in an oral culture by making knowledge the provenance of everyone who can read and write, not just the elders, those whose memories extend the furthest within the social collective (Lum, 2006). So, while the introduction of literacy changes society's power structure, it also has a profound impact on people's thinking processes

(Ong, 1982), the way they understand their relationship to the rest of the world, which is to say, how they perceive reality.

Since it is not until literacy that there is individualism, privacy, and autonomy – ways of thinking of individuals as separate not only from others in the same social collective, but also from the rest of their environment – then we can say that with writing systems come the beginnings of thinking in psychological terms. Using our present-day concept of psychological well-being, perhaps we can say that, in the second communication epoch, psychological well-being is in fact a sense of self, of being separate from the rest of society, while also belonging to the society. It is the very beginning of finding balance between belonging and separating. While there was concern in the early days of literacy in some societies that this type of separation could only be negative (Socrates worried at the time about how writing would be the downfall of civilization as he knew it), it did not concern the intelligentsia and general citizenry to the extent that new communication forms preoccupy us today. While the change from orality to literacy resulted in major social and psychological shifts, these became more pronounced with the invention of the printing press in the age of typography.

Third communication epoch – typography Marshall McLuhan, in his book *The Gutenberg Galaxy* (1962), argues that a split in human consciousness turns on the fifteenth-century invention of the printing press. In 1450s Germany, Gutenberg introduced movable metal type with the first Western printing press. Print allowed for mass production of text and eventually resulted in higher literacy rates, the possibility of faster dissemination of ideas, and, in time, made possible the Protestant Reformation, the Scientific Revolution, and the Age of Enlightenment. Elizabeth Eisenstein (1979) details this profound impact of the invention of the printing press, linking it to the major economic, social, and political movements in Europe, from the fifteenth to the nineteenth centuries.

McLuhan argued that the invention of movable type was ultimately dehumanizing because it lessened the need for people to talk face to face. It further promoted private thinking. At the same time, however, he saw it as liberating because it remade the individual as well as the collective consciousness. Print further democratized information, and ushered in a new sense of self. In the typographic epoch we see a shift from religious notions of self and nature, to secular notions. The secular world encompassed emerging democracies, scientific theories, and economic changes like the Industrial Revolution. The shift to industrial capitalism created urbanism, which led to a very different kind of social organization and spawned grand and highly influential social and psychological theories. It was the nineteenth-century Industrial Revolution and its concomitant social, economic, and political changes that brought major social theorists such as those mentioned earlier to analyze the impact of these changes on society and the individual. Writing about loneliness and alienation, as well as theorizing about the development of various components of the psyche, the id, ego, and superego, as Freud did, would not have been possible if the concept of the individual and his/her relationship to society were not so strongly recognized as important to understand. One's sense of psychological well-being in the typographic epoch would hinge on one's ability to absorb the shift in balance again between belonging and individuating. It would require successfully balancing the increasingly important individual and autonomous self within a society that is, in part because of increasingly abstract information transmission and processing through print, becoming more bureaucratic and more specialized. With increased autonomy comes an increased sense of privacy as well as a much increased possibility for loneliness and alienation.

These negative-seeming aspects of print communication are offset by what are perceived to be the positive results of print communication for cognition. In his book *Amusing Ourselves to Death*, Neil Postman (1985) wrote about the nineteenth-century typographic mind in contrast to the mind of someone who lives in the twentieth-century age of electronic media. Postman argued that, in typographic society, an individual's linear thought, critical-reasoning skills, and reading, writing, and thinking skills are well developed because print encourages those. He pointed out that in the emerging electronic age, however, we have lost much of our ability to think that way. Could it be that we were, and still are, trying to absorb another shift in balance with the rapidly emerging electronic communication environment?

Fourth communication epoch – electronic communication Electronic communication creates its own, very different, ecology in comparison with print communication. As Innis explains (1950), the nineteenth-century invention of the telephone and telegraph changed, in a relatively short amount of time, our sense of space and time as it also changed our economy. The changes in social organization made possible with electronic communication are staggering. In the early to mid-twentieth century, radio, television, and satellite communication allowed us to understand our world in completely different terms than typographic people ever could have. In the late twentieth and early twenty-first centuries, digital communication, through cell phones, the Internet, and a host of other transmission and retrieval devices, has developed so quickly that we are at once swept up in a tidal wave of technological connections, and drowning in our inability to really comprehend yet how these communication technologies are changing the way we understand the world and each other. Likewise, we are only beginning to understand how they are changing our individual selves. Are digital communication technologies separating us from others? Are they fragmenting us? Are they bringing us into closer connection with others? Are they doing all of these?

With the rapid technological changes have come a host of predictions for the future, worries about changes in our society and changes in our selves. What everyone agrees on is that change is happening. What differs is where the focus is placed and whether these changes are considered to be ultimately good, bad, or inevitable. Sherry Turkle (1984, 2004) argued that computers change how we think. She explained that young people who grow up computing and engaging with "computational objects" are absorbing more than what appears on their screens: they are learning new ways to think and to understand (2004: B26). For example, computers allow us to experiment with our identities. Online communication gives us space for identity play, calling into question whether there is such a thing as fixed identity (Barnes, 2001; Fry and Lewis, 2008) as the self becomes increasingly decentered in the new geography of cyberspace (Poster, 1990). This new geography has created for us no sense of place, according to Meyrowitz (1985), who elaborates on how cyberspace constitutes new sites for human transactions of all sorts, while altering our roles, even changing them altogether.

Our selves may also be changing in a physical sense. Kim Toffoletti (2003) argued that the body's systems are actually changing in the context of new communication technologies. She referred to the way in which our bodies – the new posthuman bodies as she called them – must engage in new ways with an external environment that is attuned to different information networks by virtue of the various different digital mobile devices we wear. Consider the communications/entertainment devices we wear on our bodies – cell phones, iPods, BlackBerries, and the like. They are part of our daily costume, extending our senses and keeping us open to signals coming in at any time. Consider also the technologies that probe our inner bodies, taking pictures of, manipulating, and assessing the inner organs and systems. MRI and CT scanning, and of course X-rays, along with the array of medical devices which are physically inserted into the body, all allow access to, and offer information about, areas previously not accessible without surgery. The outer digital communication devices combined with the inner probing devices are technologizing the body from within and without. Visual and auditory information literally surround us, creating new environments of stimulation that are of our own making, but to which we are still adjusting (Fry, 2006).

A technology that is being used more and more among lay consumers is the Global Positioning System (GPS). In some ways a continuation of the electronic collapse of space begun in the nineteenth century, GPS satellite technology is also much more. It allows us to pinpoint where in physical space we are located at any given time. Will this new digitized collapse and configuration of space, coupled with the plethora of communication devices we wear, leave us unable to be anything but connected? What becomes of our selves as separate and private individuals?

Turkle contends that the digital generation does not share the same sense of privacy, if it conceives of privacy at all, in comparison with those who grew up in the analog era. Concern about the disappearance of privacy seems to be generationally specific. Journalist Emily Nussbaum (2007) addresses the issue of Internet sites such as MySpace, Facebook, Xanga, and Flickr where young people in droves bare themselves, literally and figuratively, for all the world to see. More acutely aware than older generations of the fact that in a digitally and visually connected world there is no privacy, young people embrace lack of privacy and, with no worries for what others' knowledge about what some may call their indiscretions now might mean for their future, they exploit their lack of privacy as they recognize the new world which they both belong to and are creating. Perhaps they best recognize that they will continue to change and that they will always be exposed and connected.

This change in a sense of privacy is closely related to what has been called the disappearance of childhood (Postman, 1982). Access to all sorts of adult content on television erases the boundaries between childhood and adulthood. Since childhood was "invented" in the eighteenth century, it was taken for granted that a child had to master at least some level of reading proficiency in order to gain access to the adult world. Now a week's viewing of almost any TV talk show or drama reveals subject matter and portrays behaviors that children were once shielded from.

In this electronic communication epoch we seem to have moved into an era that can be compared with the oral communication epoch, where there was continuous face-to-face communication. Walter Ong referred to this as secondary orality, because we still read and write, but we engage more with the immediacy afforded through electronic communication devices and we rely more on images and sounds in our communications (1982: 136). Bruce Gronbeck (2006: 356) explains that, while writing restructured consciousness in the movement from the oral epoch to the literate epoch, we have now moved from the distancing psychological life of the literate universe, "back to the participatory experience of hypertext, e-mail, and the digital re-making of visual images possible with today's computers."

While notions of privacy, place, and fixed identity are challenged, and may be eroding, what it means to be an individual and part of a larger social system is changing. The balance of psychological well-being continues to shift. Increasingly it may be that psychological well-being will hinge on our ability to keep apace of the changes, to digitally or virtually connect with others in a world where the rules about what it means to be human and to communicate with humans have changed. While this world has not changed overnight, the changes have been increasing at the rate at which electronic communication technologies have been adopted. Since the turn of the twenty-first century the change has been rapid. Theories about communication and the effects of media and content have reflected the changes.

Social-science perspective

What follows is a discussion of our current state of theorizing about communication technologies and psychological well-being, from a social-science perspective. This perspective is exemplified in its basic consideration of the importance of interpersonal communication to mental health.

Communication and mental health

The health benefits of human communication are well known. Adler and Rodman begin their book *Understanding Human Communication* (2006) by discussing the needs that human communication fulfills. High on that list of needs is the need to create and maintain satisfying human relationships.

Medical researchers have identified a wide range of hazards that result from a lack of close relationships (Duck, 1992). People who lack strong relationships have two to three times the risk of early death, regardless of whether they smoke, drink alcoholic beverages, or exercise regularly. Terminal cancer strikes socially isolated people more often than those who have close personal relationships. Divorced, separated, and widowed people are five to ten times more likely to need hospitalization for mental problems than their married counterparts. Social isolation is a major risk factor contributing to coronary disease, comparable to physiological factors such as diet, cigarette smoking, obesity, and lack of physical activity (Ruberman, 1992).

There seems little doubt that communication is essential to psychological well-being. The question is whether the equation changes when that communication occurs through a technological device, rather than face to face. Social-science researchers have pursued that question only peripherally, and with the focus on media content, as can be seen in an examination of their early research.

Early research

Concerns about the impact of media are as old as the media themselves. Systematic research into these effects, however, did not begin until the 1920s, when the film industry and radio broadcasting had begun to make major inroads into American and European culture. Major research studies are generally directed toward the newest, and most popular, media. These are the media that tend to have parents, teachers, and church officials most concerned, especially in terms of their effect on young people. That is the case today, and certainly was the reason for the introduction of systematic research in the early twentieth century. However, psychological well-being was not on the radar screen for this research – the researchers were concerned with beliefs, attitudes, and, most of all, behavior. This early research was a reaction to the mediated propaganda used during the First World War. During the war, propaganda had been so blatant, and apparently so useful, both on the part of the Allies and their enemies, that people feared newspapers, but especially film and radio, had become powerful enough to "brainwash" an innocent public by influencing them in ways that they did not realize (Lasswell, 1927).

Experts in media research have identified a number of early studies that made major contributions to the understanding of media effects (Lowery and DeFleur, 1995; Rodman, 2007), and two of these provided insights that could be extrapolated to better understanding the effect of media on mental health. The first of these was the Payne Fund studies.

The Payne Fund studies Conducted in 1929, the Payne Fund studies were the first major investigations into the effects of media. At this time, children went to the movies, on average, once a week, and many experts and lay people alike were concerned that children seemed to be modeling antisocial habits from their movie viewing. Some of the Payne Fund studies were laboratory experiments. Electrodes were attached to viewers to record their skin responses and breathing patterns. These experiments suggested, among other things, that romantic and erotic scenes did not have much effect on young children. Kids apparently did not identify with such scenes, and took little notice of them. The scenes had little effect on adults, also, because adults saw them as fiction. However, the scenes with sexual content had a noticeable effect on teenagers. In fact, the researchers pointed out that sex scenes "blew the sixteen-year-olds off the graphs" (Lowery and DeFleur, 1995: 38).

Some of the Payne Fund studies used survey methods, in which young movie viewers, their parents, and their teachers were asked to recall the effects that movie viewing had on them. The results of these surveys suggested that movie viewing was harmful to a child's health (in that it disturbed sleep), contributed to an erosion of moral standards, and had a negative influence on the child's conduct. Movie fans were seen by their teachers as badly behaved when compared with their classmates who did not attend movies frequently. Heavy moviegoers also had less positive reputations, did worse in school, and were not as popular with their classmates. The surveys that asked teenagers to recall if movies had affected their behavior showed that they had "imitated the movie characters openly in beautification, mannerisms, and attempts at love-making" (Lowery and DeFleur, 1995: 40).

The Payne Fund studies, like most social-scientific studies into media effects, did not measure or attempt to define psychological well-being. The Yin and Yang of media effects, however, were already obvious. On the one hand, movies provided relaxation and entertainment during tough times. On the other hand, they provided negative models that interfered with healthy identity formation and might easily have contributed to unhealthy levels of anxiety and depression.

Seduction of the Innocent One researcher who did purport to look at the psychological effects of media, although from a decidedly unscientific point of view, was Dr. Fredric Wertham, whose studies into the effects of comic books were extremely influential in the 1950s. Wertham, a psychiatrist, was concerned about the way children became absorbed in this medium. Some children spent two or three hours a day reading comic books. Health care professionals warned that children would find it difficult to distinguish such an attractive entertainment from real life. Their fears seemed to be confirmed in 1948 when a young boy wrapped a "Superman" cape around his neck and jumped from the window of his apartment to his death (Muhlen, 1949).

Wertham argued that comic books not only presented children with a distorted and dangerous view of life, but also contributed to juvenile delinquency (1954: 82). To prove his point, he conducted a content analysis that documented the gory detail found in comics. Along with his content analysis, Wertham reported on case studies of mentally ill children whose problems he associated with their use of comic books. He told of several cases of children hanging themselves, with comic books open beside them depicting a hanging (1954: 212).

In spite of their influence, Wertham's studies were not supported by scientifically gathered research data. They gave no statistical proof, for example, that his content analysis was representative of all comics. He simply collected the most extreme and offensive examples he could find. In addition, his subjects were far from a random sample. The children were all being seen in psychiatric clinics, and many had been referred there because of extreme behavioral problems. Wertham took at face value anything these children told him, if it could be used as evidence of the harmful effects of comic books. According to researchers who later reexamined Wertham's findings, "It would have been a very dull child, indeed, who attended Dr. Wertham's clinic and did not discover very quickly that most of his or her problem behavior could be explained in terms of the comic books" (Lowery and DeFleur, 1995: 264).

Contemporary research

Investigators from the social-science perspective of media effects moved over time from a powerful-effects, to a minimal-effects, to a mixed-effects theory of media influence, continuing with an emphasis on media content as opposed to media forms. The Payne Fund and Seduction of the Innocent studies both supported the powerful-effects model, which predicted that media would have an instant and potent influence on their audiences, causing teenagers, for example, to change their self-concepts or behavior immediately following the viewing of a movie or the reading of a comic book. Later studies (see, e.g., Lazarfeld, Berelson, and Gaudet, 1948), which failed to confirm these effects, led to the minimal-effects model, which predicted that media would have little influence on behavior. Today, researchers accept a mixed-effects model, which predicts that sometimes media will have powerful effects, sometimes minimal effects, and sometimes – depending on a complex variety of contingencies – a mixture of powerful and minimal effects.

The mixed-effects model is exemplified by Wilbur Schramm's classic research into the effects of television violence on children. After years of elegant and robust research that resulted in several hundred pages of findings on increasingly specific effects, Schramm, Lyle, and Parker (1961: 13) summarized their findings as follows:

> For some children, under some conditions, some television is harmful. For other children under the same conditions, or for the same children under other conditions, it may be beneficial. For most children, under most conditions, most television is probably neither harmful nor particularly beneficial.

Cultivation theory Media effects research from the social-science perspective continued to examine beliefs, attitudes, and behaviors, and produced findings that could only peripherally provide insight into psychological well-being. George Gerbner and his associates at the University of Pennsylvania, however, took a different tack and decided to look at how media affects the audience's perception of the world, rather than their behavior. The result was cultivation theory (Gerbner *et al.*, 1994). This theory helps explain how a person's perceptions of the world are shaped and sometimes distorted by media. Cultivation theory predicts that, over time, media use will "cultivate" a particular view of the world within users. For example, Gerbner's research found that heavy television viewers had a markedly different view of reality from light viewers. Heavy viewers overestimated their chances of being involved in some type of violence, overestimated the percentage of Americans who have jobs in law enforcement, and found people, in general, to be less trustworthy than did light viewers. Gerbner dubbed this a "mean world" syndrome (Gerbner *et al.*, 1994: 30). Cultivation theory suggests that one of the primary effects of television is to give heavy viewers a perception that the world is less safe and trustworthy, and more violent, than it really is. Gerbner's findings, and cultivation theory in general, would suggest that heavy TV viewing contributes to both a warped sense of reality and unhealthy levels of anxiety.

Media addiction People spend an enormous amount of time with communication technologies. For example, according to the *Statistical Abstract of the United States 2009* (US Census Bureau, 2009), Americans annually spend the equivalent of almost five months – 3,500 hours – using some form of media.

In fact, the number of hours spent consuming media is often presented as a mental health issue. Several researchers have pointed out that the lengthy and compulsive use of media often fits the model for addiction. For example, people tend to use certain media longer than they plan to, even though longer use periods are associated with less satisfaction. This has been true in studies of television addiction (Kubey and Csikszentmihalyi, 2002), computer addiction (Orzack, 1998), Internet addiction (LaRose, Lin, and Eastin, 2003), and video game addiction (Scheeres, 2001). Others believe that media are no more addictive than any other form of pleasant pastime (Rodman, 2001).

Whether addicting or not, the number of hours spent with communication technology decreases physical activity, and the link between physical activity and psychological well-being is well established (Biddle, Fox, and Boutcher, 2003). Lack of sleep is a typical effect of popular communication media – people sleep less, for example, so they can watch more television, while the stimulation of late-night programming can make it more difficult to sleep (Schramm *et al.*, 1961). Also, television viewing is not only inherently sedentary but also associated with the consumption of unhealthy snacks, many of which are sold in television advertising (Signorielli and Staples, 1997).

Parasocial relationships The traditional mass media – books, newspapers, magazines, recordings, movies, radio, and television – are not normally used to communicate one-on-one messages in the manner of cell phones and Internet chat. And yet they sometimes do emulate interpersonal relationships, through a phenomenon known as parasocial relationships.

The term "parasocial relationships" was coined by Horton and Wohl (1956), who defined it as "a seeming face-to-face relationship between spectator and performer" (1956: 215). Whereas at first the idea of a viewer seeing a television or cinema performer as a personal friend would seem to be a form of psychosis, perhaps one that could lead to the very unhealthy behavior of celebrity stalking, several researchers have worked with the idea of parasocial relationships as a relatively common part of many people's social systems (Auter and Palmgreen, 2000; Cohen, 2001; Cole and Leets, 1999). Recently researchers have even looked at the impact, such as increased anxiety, of viewers "breaking up" with their parasocial friends when a television series is canceled (Eyal and Cohen, 2006).

Traditional media that are not used for interpersonal communication are also of concern to us here because they use up the time that would otherwise be used for interpersonal relationships. Theorists find this substitution of virtual for real relationships to be particularly dangerous, because the media relationship is an easy one whereas interpersonal relationships are difficult. Children might learn, therefore, to reject interpersonal contact in favor of the less stressful media variety.

Environmental mastery Media's effects on psychological well-being are also seen in the area of environmental mastery – the quality of being able to manage complex demands of daily life. It makes sense that mastering a complex technology like a video game could enhance self-image. Ryan *et al.* (2006), among others, have found that playing video games satisfies deep psychological needs and improves people's well-being, at least in the short term. These studies contrast with others (Anderson and Bushman, 2001; Gentile and Anderson, 2003; Healy, 1990; Setzer and Duckett, 2000) that have found a negative impact.

Still, there can be little doubt that digital media are used to manage complex tasks, and that their ability to do so provides happiness to many of their users. However, several researchers have pointed out the problems in erasing the line between work life and family/recreational/nonwork life. Chesley (2005) explains how blurring the boundaries between work and home creates problems for families, such as when a parent takes a child's worried question about a sick pet at work, or an executive is interrupted in the middle of a family dinner with a call from the workplace. This blurring of boundaries was dubbed "spillover" by Zedeck (1992), and has become a focus of considerable research. Spillover occurs in both directions, from home to work and from work to home, and can have both positive and negative effects (Frone, 2003; Roehling, Moen, and Batt, 2003). For example, Hill, Ferris, and Martinson (2003) found that telecommuters working from virtual offices have more flexibility than traditional office workers to meet both work and family needs.

Uses and gratifications One school of thought is that electronic media inundates society with too much information, and that naturally leads to unhealthy levels of anxiety. One researcher, in fact, labeled this type of anxiety "information anxiety" (Wurman, 1980). Uses and gratifications research, on the other hand, suggests that media can have the opposite effect. The uses and gratifications approach was first described in an article by Elihu Katz (1959) and has become a staple of media research (Severin and Tankard, 2000). Katz pointed out in his original article that research before that time had been primarily concerned with persuasion. Katz suggested that rather than dealing with the question of what media do *to* people, researchers might more profitably ask what people do *with* media. The point about uses and gratifications is that communication technology does, in fact, fulfill important needs for users, and the point for our purposes here is that the fulfillment of these needs would, to a reasonable point, be beneficial to psychological well-being. These needs include the following:

- Diversion/Relaxation: Media provide an escape from routine and problems.
- Relationship/Communication Utility: Media provide topics to use in conversations.
- Personal Identity: Media tell us who we are and reinforce personal values.
- Surveillance: Media provide information about things that might affect us.

Desensitization One area of media research explored desensitization – the idea that exposure to media violence will undermine feelings of concern, empathy, or sympathy that viewers might have toward victims of actual violence. Most of the early work in this area, such as that conducted by Victor B. Cline and his colleagues (1973), involved exposure to mild forms of television violence for short periods of time. These studies indicated that viewers who watched large amounts of media violence showed less physiological reactivity to violent film clips, compared with viewers who watched only small amounts, and that general physiological arousal decreased as viewers watched more violent media. Other researchers (Krafka *et al.*, 1997) found that the desensitization effect was true for sexual depictions also.

Conclusion

Communication technologies were once divided into two types: mass media such as radio and television, and interpersonal media such as the analog telephone. Mass media were used by one organization, such as a television network, to send a message out to a large and diverse audience. Interpersonal media were used for one person to communicate with one other person. Today's era of digital media has led to the convergence of mass media and interpersonal media, which is epitomized in devices such as cell phones capable of downloading television programs, and web sites such as YouTube and MySpace that allow individuals to send out what are essentially personal messages to large audiences. This makes the question of technology's effect on the beneficial aspects of human communication especially important in today's world.

For example, cell phones and instant messages keep friends in constant touch; it stands to reason that this kind of contact should encourage positive relations with others. This effect was not supported by Kraut *et al.* (1998), who looked at new Internet users and found that greater use of the Internet was associated with declines in participants' communication with family members in the household, decline in the size of their social circles, and increases in their depression and loneliness. This was true even though their sample used the Internet extensively for communication. Kraut and his colleagues called this the "Internet Paradox." A follow-up study (Kraut *et al.*, 2002), however, found that the negative effects had dissipated over time. Could it be that, within the period of time that elapsed between both studies, heavy Internet users had integrated the medium into their lives in such a way that they were able to figure out how to maintain, even reinforce, a sense of personal well-being by using it? Could it be, perhaps, that in that period of time the Internet had become much more widely used by others, and for many different tasks and connections, so that one's sense of well-being would not be affected in a negative way, but would be even positively associated with it? Or, is it that both studies reinforce our theme that communication technology sometimes helps, and sometimes hurts, psychological well-being? Yair Amichai-Hamburger and Azy Barak have more to say about the Internet and psychological well-being in Chapter 2.

Conventional wisdom tends to see the effects of communication as a straight cause–effect, with the cause being the adoption of new media forms and/or content, and the perceived effect being positive or negative depending upon one's point of view and the theoretical model deployed. We suggest here that a more reasoned point of view is that media technologies and media content have both a Yin and a Yang, in which the positive and negative effects work together, with optimal benefits seen in a golden mean of media use. We argue that it is important to look historically at changes in communication and specifically the dominant forms of communication within a given era. As communication technologies have been introduced and adopted they have created new environments for understanding the world, including the relationship between others and ourselves. We present this idea with a special sensitivity to the fact that the concept of psychological well-being has changed significantly over time, and that the development of new forms of communication media has played a major role in that change. Future historical examinations of psychological well-being and communication technologies must include Eastern cultures, which were left out of this discussion. The evolving notions of self and society, from centuries past to the present, are not the same in non-Western cultures, and a more thorough examination of those differences is necessary.

It seems social scientists and media researchers of all stripes would benefit from further investigation into the shifting sense of self and consequent shifting sense of balance that is identified as psychological well-being as communications and specifically communication technologies evolve. Such investigations might include any number of approaches, with the larger intent to understand how different modes of communication shape content and individual as well as group and public interaction. The idea is that communications mold the reality within which we live our lives. One approach in this larger endeavor is research into the way mediated communication forms replicate a sense of *presence* (see Lombard and Ditton, 1997). Presence research looks at the ways in which electronic media can create virtual environments which, through the use of an interface, can create new realities, or new ways of "being there" that produce feelings and experiences in users which are not unlike face-to-face and other more organic encounters. Another approach is detailed historical examination of media forms in different eras. For example, examining the genre of news – what it is and how it is both defined and used by audiences – and how it has changed significantly over the past two centuries (Fry, 2007). Whatever the specific angle of analysis, the field of study expands with intriguing possibilities as communication technologies develop around us at an awe-inspiring rate.

References

Adler, R. B., and Rodman, G. (2009). *Understanding Human Communication.* New York: Oxford University Press.

Amichai-Hamburger, Y. (Ed.) (2005). *The Social Net: Understanding human behavior in cyberspace.* Oxford: Oxford University Press.

Anderson, C. A., and Bushman, B. J. (2001). Effects of violent video games on aggressive behavior, aggressive cognition, aggressive affect, physiological arousal and prosocial behavior: A meta-analytic review of the scientific literature. *Psychological Science*, 12, 353–359.

Auter, P. J., and Palmgreen, P. (2000). Development and validation of a parasocial interaction measure: The audience–persona interaction scale. *Communication Research Reports*, 17, 79–89.

Barnes, S. (2001). *Online Connection: Internet interpersonal relationships.* Cresskill, NJ: Hampton Press.

Biddle, S., Fox, K., and Boutcher, S. (2003). *Physical Activity and Psychological Well-being.* Oxford: Routledge.

Brim, O. (1994). Psychological well-being. *MIDMAC Bulletin #3.* Retrieved March 9, 2007, from http://midmac.med.harvard.edu/bullet3.html

Chesley, Noelle. (2005). Blurring boundaries? Linking technology use, spillover, individual distress, and family satisfaction. *Journal of Marriage and Family*, 67, 1237–1248.

Cline, V. B., Croft, Roger G., and Courrier, S. (1973). Desensitization of children to television violence. *Journal of Personality and Social Psychology*, 27, 360–365.

Cohen, J. (2001). Defining identification: A theoretical look at the identification of audiences with media characters. *Mass Communication and Society*, 4, 245–264.

Cole, T., and Leets, L. (1999). Attachment styles and intimate television viewing: Insecurely forming relationships in a parasocial way. *Journal of Social and Personal Relationships*, 16, 495–511.

Cushman, P. (1990). Why the self is empty: Toward a historically situated psychology. *American Psychologist*, 45, 599–611.

David, P., and Johnson, M. A. (1998). The role of self in third-person effects about body image. *Journal of Communication*, 48, 57–79.

Duck, S. (1992). Staying healthy . . . with a little help from our friends? In *Human Relationships*, 2nd edn. Newbury Park, CA: Sage.

Durham, M. G., and Kellner, D. M. (Eds.) (2006). *Media and Cultural Studies: Key works.* Malden, MA: Blackwell Publishing.

Durkheim, É. (1964). *The Division of Labor in Society* (G. Simpson, Trans.). New York: Free Press.

Eisenstein, E. (1979). *The Printing Press as an Agent of Change: Communications and cultural transformations in early modern Europe*. Cambridge: Cambridge University Press.

Eyal, K., and Cohen, J. (2006). When good friends say goodbye: A para-social breakup study. *Journal of Broadcasting and Electronic Media*, 50, 502–523.

Freud, S. ([1901] 1965). *The Psychopathology of Everyday Life* (Alan Tyson, Trans.). New York: Norton.

Frone, M. R. (2003). Work–family balance. In J. C. Quick and L. E. Tetrick (Eds.), *Handbook of Occupational Health Psychology* (pp. 143–162). Washington, DC: American Psychological Association.

Fry, K. G. (2006). Advertising *and* the body to advertising *on* the body: A socio-historical look at marketing of the future. Paper presented at the XIV Economic History Congress, Helsinki, Finland, August.

(2007). Audiences for the news, the morphing genre: Virginia Tech as a case study. Paper presented at the annual meeting of the New York State Communication Association, October.

Fry, K., and Lewis, B. (2008). *Media, Myth, Religion: Identities in space, time and technology*. Cresskill, NJ: Hampton Press.

Gentile, D. A., and Anderson, C. A. (2003). Violent video games: The newest media violence hazard. In D. Gentile (Ed.), *Media Violence and Children* (pp. 131–152). Westport, CT: Praeger.

Gerbner, George, Gross, L., Morgan, M., and Signorielli, N. (1994). Living with television: The cultivation perspective. In J. Bryant and D. Zillman (Eds.), *Media Effects: Advances in theory and research* (pp. 17–41). Hillsdale, NJ: Erlbaum.

Goody, J. (1968). *Literacy in Traditional Societies*. London: Cambridge University Press.

Gronbeck, B. (2006). The orality–literacy theorems and media ecology. In Casey M. K. Lum (Ed.), *Perspectives on Culture, Technology and Communication: The media ecology tradition* (pp. 335–366). Cresskill, NJ: Hampton Press.

Havelock, E. A. (1963). *Preface to Plato*. Cambridge, MA: The Belknap Press of Harvard University Press.

Healy, J. M. (1990). *Endangered Minds*. New York: Simon and Schuster.

Hill, E. J., Ferris, M., and Martinson, V. (2003). Does it matter where you work? A comparison of how three work venues (traditional office, virtual office, and home office) influence aspects of work and personal/family life. *Journal of Vocational Behavior*, 63, 220–241.

Horkheimer, M., and Adorno, T. (2002). The culture industry: Enlightenment as mass deception. In Gunzelin Schmid Noerr (Ed.), *Dialectic of Enlightenment: Philosophical fragments* (pp. 94–136). Stanford, CA: Stanford University Press.

Horton, D., and Wohl, R. R. (1956). Mass communication and para-social interaction. *Psychiatry*, 19, 215–229.

Innis, H. (1950). *Empire and Communication*. New York: Oxford University Press.

Jung, C. (1939). *The Integration of the Personality*. New York: Farrar and Reinhart.

Katz, E. (1959). Mass communication research and the study of popular culture. *Studies in Public Communication*, 2, 1–6.
Keren, M. (2006). *Blogosphere: The new political arena*. Lanham, MD: Lexington Books.
Krafka, C., Linz, D., Donnerstein, E., and Penrod, S. (1997). Women's reactions to sexually aggressive mass media depictions. *Violence Against Women*, 3, 149–181.
Kraut, R., Kiesler, S., Boneva, B., Cummings, J., Helgeson, V., and Crawford, A. (2002). Internet paradox revisited. *Journal of Social Issues*, 58, 49–76.
Kraut, R., Patterson, M., Lundmark, V., Kiesler, S., Mukophadhyay, T., and Scherlis, W. (1998). Internet paradox: A social technology that reduces social involvement and psychological well-being? *American Psychologist*, 53, 1017–1031.
Kubey, R., and Csikszentmihalyi, M. (2002). Television addiction is no mere metaphor. *Scientific American*, February. Retrieved March 9, 2007, from www.scientificamerican.com/article.cfm?id=television-addiction-is-n-2002-02.
LaRose, R., Lin, C. A., and Eastin, Matthew S. (2003). Unregulated Internet usage: Addiction, habit, or deficient self-regulation? *Media Psychology*, 5, 225–253.
Lasswell, H. D. (1927). *Propaganda Technique in the World War*. New York: Peter Smith.
Lazarfeld, P. F., Berelson, B., and Gaudet, H. (1948). *The People's Choice: How the voter makes up his mind in a presidential election*. New York: Columbia University Press.
Lombard, M., and Ditton, T. (1997). At the heart of it all: The concept of presence. *Journal of Computer-Mediated Communication*, 3. Retrieved March 9, 2007, from http://jcmc.indiana.edu/vol3/issue2/lombard.html.
Lowery, S., and DeFleur, M. ([1983] 1995). *Milestones in Mass Communication Research*, 3rd edn. New York: Allyn and Bacon.
Lum, C. M. K. (Ed.) (2006). *Perspectives on Culture, Technology and Communication: The media ecology tradition*. Cresskill, NJ: Hampton Press.
Marx, K., and Engels, F. (1978). *The Marx–Engels Reader*, 2nd edn. (R. C. Tucker, Trans.). New York: Norton.
McLuhan, M. (1962). *The Gutenberg Galaxy*. Toronto: University of Toronto Press.
(1964). *Understanding Media: The extensions of man*. New York: McGraw-Hill.
Meyrowitz, J. (1985). *No Sense of Place*. New York: Oxford University Press.
Muhlen, N. (1949). Comic books and other horrors: Prep school for totalitarian society? *Commentary*, January, 80.
Nussbaum, E. (2007). Say everything. *New York*, February 12, 24.
Ong, W. (1982). *Orality and Literacy: The technologizing of the word*. London: Methuen.
Orzack, M. H. (1998). Computer addiction: What is it? *Psychiatric Times XV*, 8. Retrieved March 9, 2007, from www.psychiatrictimes.com/display/article/10168/54286?pageNumber=2.
Poster, M. (1990). *The Mode of Information*. Chicago: University of Chicago Press.

Postman, N. (1982). *The Disappearance of Childhood*. New York: Delacorte Press.
(1985). *Amusing Ourselves to Death*. New York: Viking Penguin.
Rodman, G. (2001). *Making Sense of Media* (pp. 296–298). Boston: Allyn and Bacon.
(2007). *Mass Media in a Changing World*, update edn. (pp. 443–454). New York: McGraw-Hill.
Roehling, P. V., Moen, P., and Batt, R. (2003). Spillover. In P. Moen (Ed.), *It's About Time: Couples and careers* (pp. 101–121). Ithaca, NY: Cornell University Press.
Rogers, C. (1980). *A Way of Being*. Boston: Houghton Mifflin.
Ruberman, R. (1992). Psychosocial influences on mortality of patients with coronary heart disease. *Journal of the American Medical Association*, 227, 559–560.
Ryan, R., Rigby, C. S., and Przybylski, A. (2006). The motivational pull of video games: A self-determination theory approach. *Motivation and Emotion*, 30, 347–363.
Ryff, C. (1995). Psychological well-being in adult life. *Current Directions in Psychological Science*, 4, 99–104.
Scheeres, J. (2001). The quest to end game addiction. *Wired*, December. Retrieved March 10, 2007, from www.wired.com/gaming/gamingreviews/news/2001/12/48479.
Schramm, W., Lyle, J., and Parker, E. (1961). *Television in the Lives of Our Children*. Palo Alto, CA: Stanford University Press.
Setzer, V. W., and Duckett, G. W. (2000). The risks to children using electronic games. Retrieved March 10, 2007, from www.ime.usp.br/~vwsetzer/video-g-risks.html.
Severin, W. J., and Tankard, J. W. (2000). *Communication Theories*, 5th edn. New York: Longman.
Signorielli, N., and Staples, J. (1997). Television and children's conceptions of nutrition. *Health Communication*, 9, 289–302.
Stephens, M. (1998). *The Rise of the Image, the Fall of the Word*. New York: Oxford University Press.
Teske, J. (2002). Cyberpsychology, human relationships, and our virtual interiors. *Zygon*, 37, 677–700.
Toffoletti, K. (2003). Bodies in stereo: Advertising at the interface. *Metro Magazine*, no. 135. EBSCO Publishing.
Turkle, S. (1984). *The Second Self: Computers and the human spirit*. New York: Simon & Schuster.
(2004). How computers change the way we think. *The Chronicle of Higher Education*, 50, B26.
US Census Bureau (2009). *Statistical Abstract of the United States 2009*. Retrieved March 16, 2009, from www.census.gov/compendia/statab.
Weber, M. (1985). *The Protestant Ethic and the Spirit of Capitalism* (Talcott Parsons, Trans.). London: Counterpoint.
Wertham, F. (1954). *Seduction of the Innocent*. New York: Rinehart.
Wurman, R. S. (1980). *Information Anxiety*. New York: Doubleday.
Zedeck, S. (1992). *Work, Families, and Organizations*. San Francisco: Jossey-Bass.

2 Internet and well-being

Yair Amichai-Hamburger and Azy Barak

The Internet is a worldwide decentralized network of computers. It started as part of the US army network and progressed to become an academic information network, and then, as it is today, a global, open communication network that affects almost every aspect of people's lives (Bargh and McKenna, 2004; Ellerman, 2007; Haythornthwaite and Hagar, 2004). Initially, the number of Internet users was small; however, the number of users is continually expanding and is estimated at more than one billion people today.[1] Another major change has come about through the ease with which the Internet may be accessed and the number of different ways this could be done. Early on, the Internet could only be accessed through a traditional, desktop computer. Today, however, people may enter cyberspace using a portable computer, cell phone, or a pocket PC, with the result that the Internet can accompany them wherever they are. In fact, the Internet has integrated so well into people's lives that, for many, it is very difficult to imagine life without it; they receive information, communicate with others, do their shopping, plan their next trip, take part in a virtual work team, and play games in a virtual community, to mention just a few common activities.

The advent of the Internet was initially regarded by many scholars as a dubious development; for example, a longitudinal study conducted by Kraut and his colleagues linked Internet use with loss of social contact, depression, and loneliness (Kraut *et al.*, 1998). It was later regarded by many as neither intrinsically good nor bad, but rather as dependent upon the use to which it is put by its content designers and its users (Amichai-Hamburger and Ben-Artzi, 2003; Hamburger and Ben-Artzi, 2000). It is important to note that Kraut's findings were later refuted by more thorough investigation conducted by his own research group (Kraut *et al.*, 2002). Another initial observation that has undergone a change in some quarters was the perception of the Internet as a distinct and homogeneous entity, rather than a complex one made up of an infinite

[1] Internet World Stats. Retrieved October 7, 2008, from www.Internetworldstats.com.

variety of surfers and multifarious services (Amichai-Hamburger, 2002, 2005). The Internet is actually a rich, complex environment. Within that environment, a possibly infinite variety of behaviors may be found. These include extremes of behavior; for example, both acts of extreme aggression and extremely selfless, altruistic behaviors may typically be observed there: this reflects the well-known phenomenon of the two faces of the Internet (Barak and King, 2000). There are many cases where the aggressive and altruistic sites are in direct contrast to one another. The Internet contains some truly horrific web sites – for example, those that encourage young people to commit suicide – while there are other sites providing support with the aim of preventing suicide. The Internet may be seen as a parallel social environment to the physical environment, in which people with differing motives and personalities utilize its endless options and possibilities.

This chapter seeks to examine how the Internet affects the psychological well-being of its users. It opens by discussing the main psychological factors influencing Internet usage and moves on to assess how the Internet affects people's psychological well-being on the interpersonal level. The next sections focus on Internet well-being on the interpersonal and group levels. The chapter concludes with some observations and thoughts about the Internet and well-being, and some speculation as to the future psychological prospects of the Internet.

The Internet in its psychological context

The Internet creates a unique psychological environment for its users. McKenna, Green, and Gleason (2002) suggested four major factors that differentiate between Internet and face-to-face (FtF) interactions: (1) greater anonymity; (2) the diminution of the importance of physical appearance; (3) greater control over the time, place, and pace of interactions; and (4) the ease of finding similar others. Several other factors – such as written, savable text as a major communication channel and lack of direct eye contact – make online interpersonal interactions even more unlike FtF contacts.

Greater anonymity

On the Internet, people can easily maintain their anonymity; for example, they can choose a false name and hide other identifying personal details. Different sites, as well as online communication tools (e.g., chat, e-mail), vary as to the degree of anonymity they allow the user to maintain. However, even when people use their real names or disclose other personal information, they still subjectively experience relative anonymity (McKenna and Bargh, 2000; McKenna *et al.*, 2002). In some cases, in what has been termed the Online Disinhibition Effect, this sense of anonymity frees people from the social norms that exist in their physical environment and facilitates the reduction of personal inhibitions (Joinson, 2007; Suler, 2004a). Generally individuals need to feel secure in their relationships with other people before they are prepared to open up and disclose personal, intimate details of themselves. A major exception to this rule is what is known as "the strangers-on-a-train phenomenon" (Rubin, 1975). This occurs when individuals share personal information with complete strangers. A process similar to this, leading to self-disclosure, is likely to occur in cyberspace (Barak and Gluck-Ofri, 2007; Bargh, McKenna, and Fitzsimons, 2002; Joinson, 2007).

Diminution of the importance of physical appearance

Physical appearance is one of the focal determinants of the way in which we are perceived and judged by others. Cialdini (1984: 148–150) suggested that attractive people have enormous social advantages; they are seen as possessing superior personality traits and intellectual capabilities, are better liked, and are more frequently helped. These benefits are bestowed due to what is termed "the halo effect." The effect, originally suggested by Thorndike (1920), occurs when one positive characteristic of a person, such as attractiveness, leads observers to believe that other positive characteristics are present (e.g., an attractive person may be perceived as honest and caring). Just as the halo effect plays in favor of the physically attractive, conversely, people who are physically unappealing will frequently find that they are judged unfavorably on unrelated aspects. Since a typical Internet social interaction is solely text-based, the physical characteristics of the participants remain unknown. This factor will be particularly significant for people with unsightly, stereotyped, or unattractive physical characteristics, or people with visible disabilities, who are likely to suffer from prejudice and discrimination in a FtF interaction. On the Internet, however, such people have an opportunity to present themselves in any way they choose (Suler, 2004b). They may hope that if and when the interaction develops to become FtF, any negative physical features will not be relevant. This freedom from their unwanted physical characteristics may help people form a new basis for their interaction with others and help them improve their self-image and self-confidence. For such people, who may have internalized a negative social perception of themselves, interaction in cyberspace may provide them with opportunities to reframe their self-image. It is important to note that this same anonymity may be a source of deception. People who feel that they can hide behind their invisibility might deceive others about their physical characteristics and describe themselves online as looking very different from their real selves, thus abusing others (e.g., in dating sites).

Greater control

One of the unique facets of Internet communication is that, for the participant, the whole encounter is taking place in an environment of his or her choosing; this may well be a source of security and comfort. The Internet user is able to "go out to meet the world," meet strangers, take part in numerous activities, communicate through asynchronous and easily controllable tools – without leaving his/her home. Internet users experience greater control in their social interactions due to the possibilities of an easy escape from contact, the fact that they can choose to remain unidentified, and a subjective sense of privacy (Madell and Muncer, 2007).

Finding similar others

The need to "belong" is listed as one of the significant needs in Maslow's (1971) hierarchy. Being a member of a group that shares your goals and interests is likely to enhance self-esteem (Tajfel and Turner, 1986). This, in itself, is believed to be very important for well-being (Branden, 1969). It is, therefore, extremely important for people to experience being members of a group which validates them and enhances their self-esteem.

The Internet is visited by many hundreds of millions of people, all of whom have varied interests. This, together with the ease of finding details of the different interest groups, makes it an exceptionally effective way of discovering like-minded others. This is particularly relevant to people who belong to stigmatized or minority groups, since they find it difficult to collaborate with similar others offline. It was found that, on the Internet, people develop group identifications sometimes even faster than is the case in offline groups (McKenna *et al.*, 2002). The existence and accessibility of almost every subject in cyberspace enable people to express aspects of their lives or personalities in ways that would be impossible elsewhere. This also encourages people to express their different interests. For many, finding a support group of similar others may help them solve personal issues and give some respite from their difficulties (Barak, Boniel-Nissim, and Suler, 2008; Tanis, 2007). In the following sections, we review some of the major ways in which the use of the Internet contributes to people's well-being.

Well-being and individual use of cyberspace

Personal experiences in cyberspace have important implications for three essential psychological issues: the Internet as a vehicle for revealing one's "real me" and one's "true self"; the reframing of individual and subjective identity; and the effects of experiencing the Internet at different life stages.

The Internet as a tool to reveal the "real me"

Rogers (1961) emphasized the concept of the "true self," representing the real self, which, for most people, remains unfulfilled. Bargh, Fitzsimons, and McKenna (2003), Bargh, McKenna, and Fitzsimons (2002), and McKenna (2007a) argued that the unique atmosphere created in cyberspace allows people to share self-relevant information in a way that they would be unwilling to do offline. Consequently, they used the concept of the "real me" based on the "true self" concept to refer to a version of the self that an individual believes to be the genuine one, but finds difficulty in expressing. It should be noted, however, that whether people reflect their "true selves" online, or just parts of those selves, as related to personal and environmental circumstances (Amichai-Hamburger, Wainapel, and Fox, 2002; Suler, 2004a), remains an unsolved debate.

Amichai-Hamburger *et al.* (2002) attempted to relate the "real me" concept to the extroversion and neuroticism personality theory (Eysenck and Eysenck, 1975). They found that among participants who were regular users of "chat," introverted and neurotic people locate their "real me" on the Internet, while extroverts and nonneurotic people locate their "real me" through traditional, FtF interactions. They suggested that the social interaction options found on the Internet (e.g., chat, forums, instant messaging) provide a solution to people who experience great difficulty in forming social contacts due to their introverted or neurotic personalities.

The concept of the "real me" enables us to understand the importance of the Internet for certain types of people; for example, those people who find that they express themselves more effectively on the Internet than through the more traditional channels of communication. This implies that, for a significant number of people, such as introverts, neurotics, the lonely, and the socially anxious, the Internet may become a very significant part of their lives and perhaps the only one through which they truly and authentically express themselves. This applies to children exploring their identities (e.g., Maczewski, 2002), disabled people behaving more freely (e.g., Bowker and Tuffin, 2002), or those with sexual identities that differ from the mainstream (e.g., Durkin, 2004; McKenna, 2001).

Reframing and experiencing identity

The Internet's secure environment may help young people to find an answer to the all-pervasive question "Who am I?" Erikson (1968) believed that a game may serve as a means through which experience is formulated in a culture that typically does not allow or encourage a search for self-identity. Turkle (1995) argued that an Internet identity game (consistent with Suler's [2002] concept of identity management in cyberspace) helps to bring about psychological maturity, which can be seen as the ability to answer the question "Who am I?" This is achieved by being able to discover different aspects of the self and experiencing flexible transitions between the various identities. Turkle believes that the Internet supplies an individual with space, warmth, safety, and understanding. This is, in fact, a similar setting to that provided by psychotherapy, so that both the Internet and the psychotherapy room may create a safe environment in which to rework elements from the past and try out different alternatives for the present and the future. In addition, the Internet supplies an environment that helps to create a "time out" for the individual and so it can serve as a moratorium, as was recommended by Erikson for adolescents in our society.

Multiplicity of selves and self-presentation One way to examine whether the multiplicity of selves offered by the net enhances Internet users' well-being is to examine to what extent this situation influences an individual's self-image. It has been suggested that individuals who perceive themselves as having a multifaceted self (e.g., a man who perceives himself as a good father, husband, worker, tennis player, and cook) are better equipped to face life stresses than those who have a more limited aspect of self (Linville, 1987). It seems that the protected environment created by the Internet generates unique opportunities for people to explore their identity and enables them to hold varying identities simultaneously. This way, cyberspace easily allows the presentation of multiple identities, in a way that is nonexistent offline. In some cases, it may even create or promote elements of therapeutic processes, resulting in more coherent and comprehensive self-identity. However, it is important to stress that, while multiple identities could be of therapeutic value, lack of ability to master and integrate a multiple self online may, in fact, have devastating effects for individuals (Amichai-Hamburger, 2005).

Personal web site and identity Launching personal web sites and blogs could be considered as a way of constructing identity. In many cases, such web sites reflect multiple personal aspects of their owners in their integrated environment. A personal web site typically includes sections on family, work, leisure activities, and other interests. It could be argued that the more numerous and varied the sections on the personal web site, the more likely it is to reflect a more significant construction of its builder's identity. Furthermore, especially in blogs, people express themselves and share with others in a projective and sincere way, so that much of their psychological identity is revealed (Barak and Hen, 2008; Marcus, Machilek, and Schütz, 2006; Preston, 2007).

The personal effects of the Internet at different life stages

Children Children as young as 4 years old have been known to use the Internet (Rideout, Vandewater, and Wartella, 2003). Therefore, it should come as no surprise that many Internet services have been designed especially for use by children and young people; for example, the Ask Jeeves search engine has its own junior version (www.askkids.com). Children can use the Internet as an important source of information, communication, learning, and entertainment. In contrast to Kraut *et al.*'s (1998) original study, which suggested that the Internet causes a reduction of social ties and depression, most of the later research (e.g., Gross, 2004; Kraut *et al.*, 2002; Valkenburg and Peter, 2007a, 2007b) found no correspondence between the amount of time spent on the Internet by children and their daily life satisfaction measures or offline social ties. It may be that since Kraut *et al.*'s early research, the social interactive environment has changed significantly. For example, during its early stages, much of the social interaction on the Internet took place between strangers. Later, however, such interactions become very common among friends and family members. Ybarra (2004), among others, did find a tendency for children with some depressive symptoms to spend more time on the Internet, but as was pointed out by Amichai-Hamburger and Ben-Artzi (2003), it is likely that this depression is not as a result of Internet use, but rather that depressed children tend to use the net. Actually, this activity might be of relieving and therapeutic value for them (Valkenburg, Peter, and Schouten, 2006).

The Internet may be perceived by children as an environment of extremes. It can be tremendously enjoyable, allowing them to express themselves freely, develop their social skills, play games, and get information and support for all kinds of issues (Boies, Cooper, and Osborne, 2004). On the other hand, they may come across many disturbing features of cyberspace (Valkenburg and Soeters, 2001). It is, therefore, important to relate to the risks and problems for children using the Internet alongside its contribution to their well-being. This is especially important as much research evidence in recent years shows that children are exposed online to numerous negative experiences, including damaging content, such as static text, pictures, video, chat, bullying, and sexual, pedophilic harassment (Allbon and Williams, 2002; Livingstone and Bober, 2005; Mitchell *et al.*, 2003, 2007; Wolak *et al.*, 2007; Ybarra and Mitchell, 2007).

It seems that child protection should be organized around three main areas: (1) legislation and law enforcement; (2) filtering software; and (3) preventive education to children on how to use the Internet wisely and safely (Dombrowski *et al.*, 2007; Fleming *et al.*, 2006; Tynes, 2007; Valcke *et al.*, 2007; Wishart *et al.*, 2007). However, the key factor behind any success is the high involvement by parents. Most young people do not tell their parents about their bad experiences on the Internet (Cole, 2001; Mitchell *et al.*, 2001). It is also the case that many parents know far less than their children about the workings of the Internet. One important step may be to increase opportunities for parents to learn about computers and the Internet, and encourage them to get involved with their children's online doings as a part of their parental responsibilities.

Adolescents Adolescence is the transition stage between childhood and adulthood and involves a search for personal identity. The physical, social, and psychological changes adolescents experience produce feelings of uncertainty in their self-definition (Feshbach and Weiner, 1991). Use of the Internet provides several types of assistance. For example, Dubow, Lovoko, and Kausch (1990) suggest that three main factors prevent adolescents from seeking medical advice: (1) the feeling that no one can help them with their specific problem; (2) the feeling that the problem is too intimate and they are not comfortable opening up and talking about it with others; (3) the fear that their problem will not be handled discreetly. On the Internet, however, while maintaining their anonymity, adolescents freely search for what they see as embarrassing information, on topics such as sexual issues; they may also approach experts anonymously (Borzekowski and Rickert, 2001; Gray and Klein, 2006; Gray *et al.*, 2005). By participating in forums that deal with typical adolescent concerns, young people may come to realize that those concerns are universal and prevalent; this may make the problems seem normative and hence less troubling. In addition to receiving advice and support from experts, users receive help from peers who take part in online forums, all of which is likely to be effective in improving their situation through an empowerment process (Barak, Hen, Boniel-Nissim, and Shapira, 2008; Becker *et al.*, 2004).

The mental pressure experienced by young people during this transition period may cause them to be particularly vulnerable emotionally. For young people aged between 15 and 24 years, suicide is one of the leading causes of death. In such situations, the anonymity granted by the Internet is not a positive factor in generating their well-being, since online such emotionally vulnerable people are likely to experience little difficulty in locating web sites that encourage them to commit suicide and even provide specific "how to" instructions (e.g., Alao *et al.*, 2006).

Adults People may be helped by the Internet to increase their personal well-being. One way to accomplish this may be by their expressing inhibited sides of their identity. Particularly for men, personal emotional exposure may be practiced with reduced inhibitions through online counseling (Rochlen *et al.*, 2004). Women are more likely to benefit from the joyful feelings of equality and independence they receive through Internet communication because of its egalitarian nature brought about through anonymity and invisibility (Suler, 2004a). The Internet may also empower women by giving them the ability to express themselves more freely in male-dominated societies, though empirical data do not consistently support this claim (Morahan-Martin, 2000, 2004). For example, they can read and write about women's issues in an open and sincere way and may access the many web sites dedicated to women's empowerment, encouraging them to reframe their self-definition. One example of this is the WORTH program, which aims to give poor women around the world the opportunity to discover their inherent strengths (www.pactworld.org/cs/worth). An early study by Hamburger and Ben-Artzi (2000) found that it is neurotic and introverted women who tend to make greater use of the social services on the Internet. The authors suggest that this is because such women have higher levels of self-reflection and a greater willingness to reach for the support that the Internet can provide. In addition, many Internet environments and tools offer women a variety of information and support related to female-relevant issues, such as breastfeeding (e.g., Velillas *et al.*, 2007), breast cancer (e.g., Owen *et al.*, 2005), sexual victimization (Jerin and Dolinsky, 2007), and pregnancy and mothering (e.g., Ley, 2007).

It is, however, still evident that women, more than men, tend to suffer from computer anxiety (Schottenbauer *et al.*, 2004) and Internet anxiety (Tsai *et al.*, 2001). It is important that the educational system tackles this issue as early as possible so that women are able to use and benefit from the Internet.

The elderly The widespread belief that old people are unable to cope with computers and would have no use for them anyway has been, sadly, internalized by many elderly people and so for many has become a self-fulfilling prophecy (O'Hara, 2004). Despite this, Eastman and Iyer (2004) report that elderly people, over the age of 65, are the fastest-growing group among Internet users.

Blit-Cohen and Litwin (2004) pointed out that use of the net by the elderly enables them to maintain a social network from their homes, since they appear ageless on the Internet and so are not immediately classified by their age and appearance with the resulting stigma attached. The elderly tend to make the greatest use of the communication and information aspects of the Internet, more specifically e-mail and searching for medical information (Fox *et al.*, 2001). Based on numerous observations in a variety of cultures and settings, researchers (e.g., Carpenter and Buday, 2007; Fozard and Kearns, 2007; Fuglsang, 2005; Karavidas *et al.*, 2005; McMellon and Schiffman, 2002; Xie, 2007) suggest that the computer helps the elderly to counter four major challenges with which they are faced: loneliness, boredom, learned helplessness, and a reduction in their mental abilities. The Internet enables the elderly to establish communities with no geographical or physical limitations (see, e.g., www.seniornet.com). This may prove an extremely important way in which to maintain their relationships with friends and family, and make new acquaintances. This may prove particularly important should they reach a stage of physical disability. In this way, the Internet may serve to give older people important feelings of control over their lives. The elderly may access the Internet through components developed for the sight-impaired: for example, large fonts and vocal browsers (Rau and Hsu, 2005). Several field intervention projects have shown that older people's training and use of the Internet can indeed empower them and contribute to their increased social contact and reduced feelings of loneliness and depression (e.g., Shapira, Barak, and Gal, 2007; White *et al.*, 2002). It is important to assist the elderly community to expand their knowledge of computers and Internet use and remove psychological and technical barriers so that they can gain from the Internet to promote their well-being (Adams *et al.*, 2005; Dunning, 2004; Meischke *et al.*, 2005).

Well-being and interpersonal interactions in cyberspace

The Internet has a very significant impact on people's well-being in terms of their interpersonal contacts and relationships. Below we will discuss the impact of online interactions on well-being by exploring the topics of interpersonal relationships, the patient–physician relationship, online psychotherapy and counseling, and online support groups.

Interpersonal relationships

Larson (1978) pointed out that people who have more friends or spend more time with their friends are happier. Argyle and Furnham (1983) found that the spouse relationship is the greatest source of life satisfaction, with close relatives and friends next. When it comes to marital status, married people score higher on life satisfaction as compared with those who are unmarried; those in widowhood scored the lowest (Nettle, 2005). Baumeister and Leary (1995) argued that the need for close personal relationships is a fundamental and a highly pervasive motivation. In their study, they found that a lack of attachment is linked to a variety of ill effects on adjustment, health, and well-being. If belongingness is a basic need, the Internet constitutes a significant social environment through which this basic need may be fulfilled. The Internet provides a convenient means through which to build and maintain relationships, for both friendship or romantic purposes.

Friendship and intimate relationships involve mutual trust. When we trust someone, we are likely to feel comfortable with him or her and disclose intimate details about ourselves. Since self-disclosure is usually a reciprocal process, within a dyad disclosure, it often serves to strengthen the ties that bind people in friendship-based relationships or romantic ones (Jourard, 1971). While this is usually a lengthy process, as mentioned earlier, a far speedier process toward disclosure may take place when two strangers meet with no expectation of ever meeting again (Ben-Ze'ev, 2005; see also Rubin, 1975, "the strangers-on-a-train phenomenon"). A similar process leading to fast self-disclosure frequently occurs in cyberspace, due to the protected anonymous environment it supplies (McKenna *et al.*, 2002) and other factors related to the online disinhibition effect (Suler, 2004a). Indeed, accelerated and increased self-disclosure is evidently typical in virtual interpersonal relationships (e.g., Chiou, 2006; Cho, 2007; Dietz-Uhler *et al.*, 2005; Joinson and Paine, 2007; Qian and Scott, 2007) as well as reciprocity of self-disclosure (Barak and Gluck-Ofri, 2007).

Ben-Ze'ev (2005) argued that this unique anonymous psychological environment allows introductions and interaction without gating features. Since online communication is the only way to learn about one another, this, together with the ease of finding like-minded people, encourages self-disclosure. A large number of studies have investigated the formation and maintenance of interpersonal relationships in cyberspace via chat rooms, forums, social networks, dating sites, and many more online options. Much of the research reports that most people find these relationships satisfying and enjoyable and that they lead to a broadening of social circles. This runs counter to some of the earlier contentions (e.g., Kraut *et al.*, 1998) that the Internet limits or restricts FtF contacts and leads to social isolation. These studies include reports of people sharing and disclosing online (Henderson and Gilding, 2004; McKenna and Bargh, 1998; Parks and Floyd, 1996), empowered to make more friends (Tidwell and Walther, 2002), though some patterns of relationships might be different online from offline (Chan and Cheng, 2004; Cheng *et al.*, 2006). Moreover, growing research shows that online relationships contribute to people's well-being (Valkenburg and Peter, 2007a; Valkenburg *et al.*, 2006), apparently by allowing more openness and self-disclosure, ventilation, and social support. It is important to note that social relationships may be established and kept up through various online vehicles, including instant messaging (Bryant *et al.*, 2006; Hu *et al.*, 2004), chat (McKenna *et al.*, 2002), forums (Barak and Gluck-Ofri, 2007), blogs (Stefanone and Jang, 2008), or social networks (Ellison *et al.*, 2007; Hlebec *et al.*, 2006).

Internet interaction may provide opportunities to make contacts and preliminary acquaintances, which may later be followed by FtF meetings. Baker (2002, 2005, 2008), Hardey (2002), and Whitty and Carr (2006) reported that many Internet romantic encounters progress gradually to include phone conversations, exchange of photographs, exchange of pictures and letters, and finally meeting in person. A study by McKenna *et al.* (2002) found that Internet-mediated dating, and romantic relationships in particular, produced positive effects on psychological well-being. Sixty-eight percent of participants reported an increase in their social circle. Forty-seven percent reported a decrease in their feelings of loneliness. Twenty-five percent of the survey's respondents indicated that the relationship decreased their feelings of depression.

Fantasy in cyberspace One of the aspects of online, text-based romantic relationships is the element of fantasy that they are likely to engender, as one Internet surfer wrote:

> I've been chatting with a girl, the most unbelievable person I've ever met. I mean, this girl is tall and has short brown hair – which happens to be my favorite hair style. On top of that, she has the biggest blue eyes you can ever find. I haven't really seen a picture, but she described it in fine detail . . . Now, on top of all that, she's smart and she's getting a Ph.D. (Mileham, 2003: 71)

These types of fantasies, not necessarily related to romantic relationships, are characteristic of virtual communication, where ambiguity and uncertainty prevail. Apparently, these fantasies stem from, or are related to, personal needs and wishful thinking, and project individuals' personality-related desires (Barak, 2007b; Turkle, 2004). It is important to note that when people create cyber-romantic relationships with such "perfect" partners, their offline alternatives tend to compare unfavorably (Ben-Ze'ev, 2005; Young *et al.*, 2000) and lead to a diminution in the energy and interest a person invests in offline partners. Virtual romantic relationships may be satisfying and reinforcing for some, especially those who refrain from physical meetings due to personal conditions, such as disability, shyness, and/or lack of self-confidence. A FtF meeting following on from virtual romance may lead to frustration and anger, as illustrated by Goldman's (2003) report of a "handsome man" called Tyrone, beloved of many female Internet users. Tyrone even sent his photograph to anyone who requested it, and was sent presents and money by many of the women, each believing that she was the only love in his life. This continued until one of his admirers tracked him down and discovered that Tyrone was actually a married woman. The police were called and Tyrone was later hospitalized following a nervous breakdown. Similar stories appear to be fairly common in cyberspace.

A social compensatory tool As we discussed earlier, the Internet can serve as a compensatory tool (Hamburger and Ben-Artzi, 2000). This factor may be very significant in online romantic relationships as well as in other social contacts. Shy people who typically freeze during FtF interactions frequently find that, in online romantic encounters, they are able to express themselves as they would wish. It is the protection the Internet provides – due to its easy escape potential, lack of eye contact, and invisibility – which "allows" them to express another side of their personality and open up to their partner online (Amichai-Hamburger, 2005; Suler, 2004a). Such people find it easier to disclose their inner selves and intimate thoughts to others through the Internet rather than in FtF meetings, and they will be more likely to form close relationships online (McKenna *et al.*, 2002). They also tend to bring their virtual relationships into their offline lives.

Online dating Online dating is the modern version of traditional matchmaking, where the matchmaker's role is replaced by sophisticated Internet technologies. Many of the dating sites are sectarian, seeking to target and service a particular group, for example vegetarians, Hindus, or academics. Other dating sites are intended for the general population. Popular dating sites claim millions of members and numerous successful partnerships. Generally, people will choose their prospective date based on personal information held by the web site, pictures, as well as individual writings presenting the self and expectations for a partner and a relationship (Baker, 2005; Lawson and Leck, 2006; McKenna, 2007b; Whitty and Carr, 2006). Individuals can freely choose whether or not to respond to invitations they receive from potential partners without feeling embarrassed or guilty. Ordinarily, the first stage of interaction occurs in cyberspace; partners choose to meet only if they feel comfortable in the preliminary relationship. Serious problems related to online dating sites have to do with deception, impersonation, and exploitation.

Social ties and reduced loneliness As mentioned earlier, although it was initially suggested that Internet use causes people to become isolated and lonely (Kraut *et al.*, 1998), most subsequent studies demonstrated that in fact many lonely, anxious, shy, and depressed people found the Internet to be an environment that allowed them to interact with others and overcome their limitations (e.g., Amichai-Hamburger and Ben-Artzi, 2003; Kraut *et al.*, 2002; Valkenburg and Peter, 2007b). In this sense, using cyberspace for social purposes can clearly be seen as enhancing well-being.

Generalizing from an online to an offline relationship As mentioned above, the Internet can serve as a compensating tool for people with social inhibitions (Amichai-Hamburger, 2005). Although many people aim to transfer their online relationship offline (e.g., McKenna *et al.*, 2002), for some people the social skills acquired online might not be generalized to offline contacts. Amichai-Hamburger and Furnham (2007) suggested an approach to deal with this problem. This approach described a gradual process, whereby the level of exposure in the interaction is increased from one stage to another. The socially anxious person moves from one stage to another when he or she feels ready for it. The main stages proposed in this process include initially text only, where basic interpersonal interaction takes place online based on written text; a user remains at this stage until his/her social anxiety declines and he/she is ready to move to the next stage. Text and video constitute the second stage, where the interaction continues by text and, in addition, people are able to see one another through live webcam. In the third stage, audio and video are used, where participants still stay in their secure environments, but interaction is based on both vision and sound. The final stage is actual FtF interaction. Changes from one stage to the next occur only when a person is ready in terms of a low level of anxiety, so that the likelihood of generalizing online social skills to relationships offline is significantly enhanced. This approach demonstrates the ability of the Internet to serve not only as an environment that allows people with social inhibitions to interact, but also as one that provides tools and opportunities to transfer social skills from online to offline relationships.

Improving the patient–physician relationship

An increasing number of people are turning to the Internet for health-related information. Baker *et al.* (2003) found in a survey that approximately 40 percent of respondents with Internet access reported using the Internet to look for advice or information about health. Six percent reported using e-mail to contact a physician or other health care professional. About one-third of those using the Internet for information on a physical condition reported that their use of the Internet affected their decisions. However, very few reported that it impacted measurable health care utilization; 94 percent said that Internet use had no effect on the number of visits they made to their physician, and only slightly fewer, 93 percent, reported that the Internet had had no effect on the number of telephone contacts they had with their physician. Five percent or less reported using the Internet to obtain prescriptions or purchase pharmaceutical products. Ybarra and Suman (2006) and Fox *et al.* (2001) pointed out that almost 48 percent of the people looking for health information said the data they found online helped them take better care of themselves. A large-scale study by Bundorf *et al.* (2006) found that health information on the Internet is used particularly by populations with serious health needs and those facing significant barriers in accessing health care. Ybarra and Suman (2006) reported that access to reliable health information online has been linked to reduced anxiety, increased feelings of self-efficacy, and a decrease in utilization of ambulatory care. Ayers and Kronenfeld (2007) demonstrated the importance of seeking and utilizing online health information to chronically ill people. Allwes and Popovich (2007) showed how people feel empowered by their ability to find and use health information. In this regard, it is important to note that online health information might be of special importance for older people due to their limitations (Flynn *et al.*, 2006; Hardt and Hollis-Sawyer, 2007).

In recent years, the medical profession has been encouraging greater understanding by patients of their medical conditions and greater participation in the decision-making process pertaining to this. The advent of the Internet has both accelerated and altered this process. The accessibility of information via the Internet has created a shift in the patient–physician relationship; while in the past, the physician possessed knowledge of which patients were ignorant, today many people learn about conditions before they seek medical intervention. This shift may lead to changes in the way in which medical decisions are taken. For example, Podolsky (1998) argued that the Internet has had a "leveling effect" on the access to information and, subsequently, on the patient–physician relationship which, in turn, may lead to defensive attitudes on the part of the medical profession. Others maintain that, in fact, the decision-making process may improve if efforts are made to share the burden of responsibility and knowledge. Moreover, further benefits may arise from physicians who assist patients in the information-gathering process (Charles, Gafnil, and Whelan, 1997). Relationships between patients and health professionals in general have altered (McMullan, 2006) and interestingly, use of medical online searches has in many cases changed the terminology patients use with physicians (Jucks and Bromme, 2007). Gerber and Eiser (2001) suggested that health providers should refrain from feeling frustrated that their role as the source of information has been challenged, or they run the risk of losing patients, since highly motivated patients wish to be active participants in their health care. Physicians should direct their patients to valid web sites and show them how to evaluate the credibility of information.

Internet-delivered psychotherapy and counseling The idea of providing psychotherapy and counseling through the Internet – e-therapy – until recently might have been considered fanciful. However, advanced Internet technology has changed the tools and options available to therapists and clients to offer and receive mental health interventions online (Suler, 2008). Today, thousands of professional therapists around the world offer psychotherapeutic interventions online by a variety of methods, which include taking advantage of remote access and communication, asynchronicity, textuality and hypertextuality, and use of rich media (Anthony, 2006; Barak, 2004; Manhal-Baugus, 2001; Ybarra *et al.*, 2005). Internet-assisted (or -delivered) therapy may be classified into four main categories: (1) Web-based therapy, where patients enter a site – which could be static or interactive, closed to authorized people or open to all, free or for a fee. This type of site is based on self-guided therapy, and follows a particular therapeutic intervention model (e.g., Cognitive Behavior Therapy). By following the steps of the program detailed at the site, clients go through a therapeutic process. The program monitors the clients, who are requested to fill out online questionnaires, take part in online and offline activities, and report back details of their experiences. To receive additional help, the site may provide further channels through which clients may communicate online with therapists – synchronously or asynchronously, sometimes also by phone. Web-based therapy is offered for various problem areas, such as depression, social anxiety, insomnia, and many more. Reviews of web-based therapy, drawing on both empirical studies and accumulated practice, provide much support for this approach (e.g., Andersson, 2006; Carlbring and Andersson, 2006; Etter, 2006; Strecher, 2007; Weinstein, 2006). (2) Online therapy (or e-therapy), where therapist and clients communicate and conduct counseling and therapeutic conversations through an online communication tool. Communication could take place synchronously (through chat or instant messaging) or asynchronously (through e-mail), individually or in groups (through chat room or forum), and by use of audio (e.g., through Skype) and/or video (using webcams). These therapeutic conversations may be assisted by directing clients to online readings, test-taking, or other online activities. Various psychological approaches may be used, including psychodynamic, cognitive, and behavioral. Literature reviews of online therapy (e.g., Grohol, 2004; Mallen *et al.*, 2005; Skinner and Zack, 2004), based on empirical research and intensive practice, show that, despite the physical distance, invisibility, and mostly textual-based communication, such psychotherapeutic interventions are quite successful. (3) Online software, through which clients are being treated by a robotic therapist (e.g., Epstein and Klinkenberg, 2001) or by the assistance of an advanced program (e.g., virtual reality; Riva *et al.*, 2007). (4) Hybrid therapy, where integration of offline therapy and online activities and communications are offered (Fenichel *et al.*, 2002; Ritterband *et al.*, 2003). This approach allows a therapist to utilize online options, combined with regular FtF meetings, so therapy is enriched and accelerated. In this process, clients take part in various online activities, including publishing a blog, participating in an online support group, corresponding with the therapist between FtF sessions and/or after termination of therapy, and visiting sites for therapy-related readings.

Internet-assisted therapy is not flawless. There are a number of professional, technical, ethical, legal, and practical issues that the therapist has to deal with (e.g., Kraus, 2004; Manhal-Baugus, 2001; Ybarra *et al.*, 2005). In recent years, however, many of these obstacles have been overcome through technological advancements, establishment of specific ethical guidelines, and therapist education and training. A recent meta-analysis (Barak, Hen, Boniel-Nissim, and Shapira, 2008) found that, on average, online psychotherapeutic interventions are just as effective as traditional, offline interventions. Apparently, this innovative possibility to receive professional help significantly contributes to people's well-being.

Several online support services throughout the world offer free counseling and support to anonymous people in distress. This approach is of special importance for suicidal people who refrain from referring to traditional therapy. One outstanding example of this is SAHAR, an online support service dedicated to helping people suffering severe emotional distress and those contemplating suicide (Barak, 2007a).

Well-being and cyberspace groups

People have a basic need to belong and affiliate with others (Watson and Johnson, 1972). Belonging to a group may have implications for an individual's self-definition and self-esteem (Brown, 1978). When the group becomes part of their self-definition, any pressure on the group's values is likely to impact their self-esteem (Tajfel, 1978; Zander *et al.*, 1960). This human need is as important online as it is offline. People use online group membership to realize these needs in the same way as they do their offline group membership.

A group and a community are generally considered as separate entities, with our community having a more comprehensive and deeper impact on our identity. In the online world, these differences are sometimes blurred. Researchers in this area (e.g., Brandon and Hillingshead, 2007; McKenna and Seidman, 2005; Wellman *et al.*, 1996) refer to online groups as entities that give their members a sense of belonging, companionship, affiliation, and social and emotional support. Rheingold (2000) suggested that a community is a collective of people who together endeavor to achieve collective good. Three examples of the collective good stand behind a typical virtual community: (1) social contacts; (2) acquisition of knowledge; and (3) identification and emotional attachment. Becoming a member of an Internet group is usually the result of a positive choice (which is not always the case in an offline group). It is, therefore, quite natural that online community members report enjoyable experiences and fun (Wasko and Faraj, 2000). An associated recent development in cyberspace has been that of social networks that combine advanced personal space (including personal blogs) and interpersonal, group-like communication (Boyd and Ellison, 2008; Haythornthwaite, 2007). Millions of people have joined major social networks (e.g., MySpace) and intensively engage in social interactions more than ever before.

Below we discuss five aspects of online groups which are pertinent to well-being: (1) stigmatized groups; (2) support groups versus treatment groups; (3) online contact between rival groups; (4) religious groups; and (5) solving the conflict between belonging to a group and individuality.

Stigmatized groups

Online groups exist on almost every possible topic. This is of particular use to people who have an unusual or esoteric interest that they find difficult to pursue offline because of the challenge of finding similar others. On the Internet, with hundreds of millions of surfers from all over the world, it is much more likely that this problem will not arise. Moreover, on the Internet, the locating of others like yourself is very straightforward. This has important implications for people who belong to groups with a negative stigma. This stigma can be due to physical features (for example, deformity or scarring), skin color, physical or mental illness, political opinion, or any other tendency that is perceived as negative or unattractive by society. In the offline world, belonging to such groups is likely to lower the self-esteem of members (Link *et al.*, 2001). Social rejection may be a persistent source of social stress and may increase feelings of self-deprecation (Wright *et al.*, 2000). Offline, for those belonging to certain types of such groups, the ability to conceal their membership may be seen as an advantage, but over time, for several major reasons, this may well become a disadvantage: (1) It is difficult to find similar others as they, too, are likely to be hiding their stigmatized identity. (2) As a group member, you are very likely to be subjected to hearing negative observations about your stigmatized group, since people are unaware that you belong to this group and so will not restrain their comments; this is likely to lead to a diminution in self-esteem (McKenna and Bargh, 1998). (3) Another blow to one's self-image may arise from the difficulties involved in locating similar others. Online, it is possible to openly belong to a group and find similar others without such dangers.

The group on the net is likely to be lively and fulfilling, and participants will frequently receive positive reinforcement, encouragement, and support. All of this may well have such a significant impact on their self-image and self-esteem that members may feel ready to reveal their secret, stigmatized identity to family and friends (McKenna and Bargh, 1998).

Support groups and therapy groups

Emotional and informational support enhance psychological well-being (Thoits, 1986; Uchino *et al.*, 1996). However, such restrictions as time, geographical location, physical debilitation, and the difficulties involved in opening up to others on intimate subjects may limit the success of offline support groups. However, conversely, these are the factors that lead to the success of such support groups online.

Online support groups cover a vast range of topics; included among them are: rare skin cancer (Lamberg, 1997), bone marrow transplant patients (Farnham *et al.*, 2002), hearing difficulties (Cummings *et al.*, 2002), foreign students (Chang *et al.*, 2001), disabilities (Anderberg, 2007), dental anxiety (Buchanan and Coulson, 2007), depression (Houston *et al.*, 2002), infertility (Epstein *et al.*, 2002), cancer of various types (Blank and Adams-Blodnieks, 2007; Lieberman and Goldstein, 2006), older adults (Wright *et al.*, 2000), alcoholism, HIV (Davison *et al.*, 2000), and suicide prevention (Barak and Dolev-Cohen, 2006). Support groups are, however, open to abuse, for example, by people impersonating and inventing personal horror stories and consequently receiving much emotional support (Barak, 2005; Feldman, 2000). This type of deception, should it become known, may damage the credibility of the group. Van Uden-Kraan *et al.* (2008) pointed out that the potential disadvantages of online support groups, for example, having low-quality messages and flaming (since no control over the information exchange exists), were present only in a minor proportion of the postings found on such sites. Support groups cannot replace psychotherapy or professional counseling, but can offer much emotional relief that complements (online or offline) psychotherapeutic interventions, provides personal empowerment, and improves the well-being of patients (Barak, Boniel-Nissim, and Suler, 2008; Owen *et al.*, 2004).

Online group therapy Both support groups and therapeutic groups are available on the Internet. Several factors differentiate between them: (1) Support groups usually do not have a professional supervisor or leader, while treatment groups must be run by a professional therapist, expert in group interventions. (2) In contrast to support groups, therapy groups have an explicit goal to bring about substantial change in patients' emotions and behaviors. (3) Support groups do not follow professional treatment principles and protocols; they are very flexible and adapt to the spontaneous wishes of members. Therapy groups follow certain structured models, consistent with specific, predetermined therapeutic approaches. (4) Support groups are usually open to anyone who wishes to join, while applicants to a therapy group undergo a screening procedure to assess their suitability to the group. (5) Taking part in online support groups is usually free, but participation in a therapy group comes with a fee.

Although online therapy is less popular than online support groups, there are an increasing number of studies that demonstrate the effectiveness of this type of therapy. Barak and Wander-Schwartz (2000) found that online and offline therapeutic groups showed the same degree of positive improvement with regard to participants' self-image, social relationships, and well-being. The online group, however, reported higher levels of aggression than the FtF group. Participants in both therapy groups expressed similar general satisfaction with their respective group therapy. Rotondi *et al.* (2005) examined online group therapy for people with schizophrenia and their families and showed that, after three months, participants with schizophrenia reported lower perceived stress and showed a trend for a higher perceived level of social support. Celio *et al.* (2000) examined the effectiveness of online group treatment for people with body dissatisfaction and disordered eating behaviors/attitudes. People who took part in the online treatment reported reductions in weight/shape concerns and disordered eating attitudes. Disordered behaviors were also reduced as compared with the results of the control group. Hopps *et al.* (2003) reported on an online therapy group for disabled people suffering from social isolation. Reviews by Bellafiore *et al.* (2004) and Colòn and Friedman (2003) highlighted the advantages of conducting group therapy online, though pointed at some difficulties, such as flaming and misunderstandings. Obviously though, the ease of access, invisibility, and savable textuality encourage people in need to join group therapy through the Internet. Despite technological difficulties that characterize this endeavor (e.g., misunderstandings in the group due to lack of nonverbal cues), it seems that growing evidence and promising emerging technologies will accelerate this therapeutic approach (Golkaramnay *et al.*, 2007; Heinicke *et al.*, 2007; Marziali, 2006).

Online contact between rival groups

Intergroup conflict is sadly part of our existence. Such conflicts exist around the globe originating through differences, for example, in beliefs, religion, race, and culture. In what became the contact hypothesis, Allport (1954) suggested that in order for a contact to create a positive effect, a set of conditions must be met. The four most important criteria are:

1. *Equal status*: Between the people who take part in the contact.
2. *Cooperation*: Between the rival groups toward a superordinate goal. This is likely to reduce the stereotypes.
3. *Intimate contact*: The contact has to create a context where the participants can really learn to know the other side. If the contact is restricted to an artificial level, stereotypes are likely to remain unchanged.
4. *Institutional support and willingness to participate*: Having the backup of the authorities is likely to create positive expectations from the

contact and build positive social norms enhancing a positive contact. Participants must contribute to the contact of their own volition and not because they have been compelled to do so (Allport, 1954; Amir, 1969, 1976).

Amichai-Hamburger and McKenna (2006) suggested that the traditional FtF contact suffers from three main obstacles: (1) Practicality – if it is to be carried out according to Allport's conditions, the contact has to contend with logistical stipulations that in many cases are difficult to achieve. (2) Anxiety – intergroup anxiety is the result of the anticipation of negative reactions during the intergroup encounter (Stephan and Stephan, 1996, 2001). Wilder (1993) pointed out that when they are in a state of anxiety, group members are likely to ignore any disconfirming information supplied in the contact context. Under such conditions, when a member of the outgroup behaves in a positive manner that contradicts the expectations of the other side, members of the ingroup do not alter their opinions and only recall the outgroup as behaving in a manner consistent with their negative perception. And (3) generalization – one of the greatest challenges to the contact hypothesis is the issue of whether or not the results of a positive contact with a member of the outgroup will be generalized further (Hamburger, 1994).

The Internet may be the key to providing the conditions required by the contact hypothesis mentioned above.

Equal status: Many of the cues on which individuals typically rely to gauge the internal and external status of others are not in evidence online.

Cooperation toward superordinate goals: Virtual workgroups have proved an effective tool worldwide (Galegher and Kraut, 1994). This is especially likely to be the case when the technology utilized provides the capability for information-rich communication (Bell and Kozlowski, 2002).

Institutional support and willingness to participate: This might well be easier to achieve online since institutions may feel that participating in an Internet contact may be seen as taking less of a risk than an FtF contact (Bargh and McKenna, 2004). This, too, may make it easier for group members to volunteer to participate and for leaders to support such a meeting.

Intimacy: Cook (1962: 75) stressed the importance of the "acquaintance potential," or the opportunity provided by the situation for the contact participants to get to know each other. One of the major advantages of Internet interactions over FtF interactions is the general tendency for individuals to engage in greater self-disclosure and more intimate exchanges there. Interactions online tend to become "more than skin deep" and to do so quite quickly (e.g., McKenna *et al.*, 2002; Walther, 1996).

Anxiety: Many of the situational factors that can foster feelings of anxiety in social situations (e.g., having to respond on the spot, feeling under visual scrutiny) are absent in online interactions because participants have more control over how they present themselves and their views online (e.g., being able to edit one's comments before presenting them). People may feel better able to express themselves and feel more at ease with their online partners than they would if they were interacting in person.

Generalization from the contact: The online environment may be said to heighten the perception of the individual members as representative of their disparate groups, while simultaneously fostering feelings of kinship and attachment to the "new group" composed of all members taking part in the exercise. There are a number of means by which the first can be achieved. For instance, all members may be given pseudonyms that are evocative of the group they represent. These recommendations have been incorporated into the Net Intergroup Contact (NIC) platform (Amichai-Hamburger, 2008).

Religious groups on the net

Religion is one of the main factors behind life satisfaction and has been found to benefit people by providing them with, most importantly, meaning in life and social support (Argyle, 1987). Myers (1992), summarizing the results of many surveys carried out both in Europe and the USA, pointed out that religious people report feeling happier and being more satisfied with life as compared with nonreligious people.

According to Hoover *et al.* (2004), nearly one-third of American adults who use the Internet in the USA use it for faith-related activities. Even in fantasy games, people fulfill their religious duties. Grossman (2007) pointed out that online, in the second life community (an online virtual world which attempts to create a total environment for its users), people have constructed places of worship for several of the world's major religions. Religious leaders in second life estimate that one thousand avatars teleport into churches, synagogues, or mosques on a regular basis. Hundreds more list themselves as Buddhist, pagan, or as belonging to other groups.

Campbell (2004) suggested that online religious communities raise a significant challenge for religious culture. A survey of cultural changes in media, community, and religion uncovered similar structural shifts, from hierarchical structures to more open, dynamic relationship patterns in society. Examining these shifts has helped to explain why cyber-religion and online religious communities have become an emergent phenomenon. It has been argued that the Internet has thrived because it has surfaced in a cultural landscape that promotes fluid yet controlled relationships over tightly bound hierarchies. Religious online communities are seen as expressions of these changes and challenge traditional religious definitions of community.

McKenna and West (2007) found that people in online religious communities receive similar benefits as those participating in traditional offline communities. These benefits are varied and include, for example, a sense of connection with others, increased social support, and sense of purpose in life. Also interesting was the fact that a significant number of online users who engaged in online interactive religious forums did not belong to an offline religious community.

Solving the conflict between belonging to a group and individuality

According to the optimal distinctiveness theory of Brewer (1991), people are driven by two contradicting motives: the need to express individuality and the need to belong to a large, significant group. Brewer differentiated between collective identity – the belonging to a large group – and the interpersonal identity – relating to others on an individual basis. These needs contradict one another: satisfying one means evoking the need to satisfy the other. People strive to belong to a group that satisfies both needs in the most optimal way (see, also, Brewer and Gardner, 1996).

Bettencourt and Sheldon (2001) attempted to resolve the dichotomy of Brewer's (1991) motive contradiction by introducing the social role. According to Biddle (1979: 9), the social role refers to "a behavioral repertoire, characteristic of a person or a position; a set of standards, descriptions, norms, or concepts held for the behaviors of a person or social position." Social roles, by definition, involve an interaction with group members. As role holders contribute to group goals through the fulfillment of their roles, they become more connected to the group. This is especially true when they fit their roles well, that is, when their social role is consistent with their core skills, leading them to feel autonomy and self-expression. Thus, the seemingly contradicting motives of autonomy and relatedness may be mutually achieved when the individual performs social roles that fit his/her ability and characteristics. This argument was proved in a series of studies (Bettencourt and Sheldon, 2001). Amichai-Hamburger (2005) pointed out that, despite the seemingly optimistic approach of Bettencourt and Sheldon, it is important to stress that, in many cases, offline groups demand that individuals take on roles not necessarily of their choosing. Those responsible for allocating these roles will inevitably not take into account the skills, abilities, and motives of the individual, but will rather look to the needs of the group. In addition, roles are frequently allocated according to the social role to which the group is aspiring. In such a case, an individual may be chosen to perform a certain role according to the general perception that is held of him or her. This perception may not be in keeping with who he or she really is, or may be outdated, and, therefore, the fit between personality and social roles will not be accurate. This may be the case even when the group attempts to choose the best person for the role.

In the Internet arena, it may be easier to find social roles that allow individuals their self-expression and even their self-actualization, since the net provides surfers with more freedom to choose the groups and the social roles that suit them. It is also the case that the number and variety of groups are far greater than in the offline world; it is frequently far easier to find such groups on the Internet than it is offline. The act of compelling someone to take on a role does not exist on the Internet, since any attempt to do so will lead the person to disagree or express alternatives and, should this not be acceptable, he or she can leave without any social consequences. It seems that the Internet creates an environment where the conflict between the need to relate and the need for autonomy is very limited and, in fact, the Internet is able to sustain situations in which both needs are mutually fulfilled. In this case, there are positive implications to be considered that pertain to individual well-being.

Some concluding ideas

This chapter has discussed the Internet environment and its significant effect on our well-being on the personal, interpersonal, and group levels. This, however, is certainly not the whole picture. In fact, the distinctions made between the different levels (personal, interpersonal, and group), while useful as methodological tools, are actually an oversimplification of reality. This is because a web site may well operate on more than one level. One example of this is Baby Carelink, a communication technology which empowers parents of premature infants. It is designed to support communication, collaboration, and consultation among doctors and patients where necessary over great distances (Safran, 2003). Below we briefly discuss the following: well-being on a wider scale; closing the digital gap; addiction; and enhancing Internet abilities.

Well-being on a wider scale

The Internet is likely to have an influence on our well-being in a large number of ways, many of which are out of the range of what is considered to be the scope of the field of psychology. Despite this, two such areas, which we believe have the potential to make vast improvements in people's well-being, are discussed below. These are e-governance and the formation of an entire Internet global village.

Since the sixties, there has been a gradual decline in the involvement of citizens in both local and national politics. One major reason for this is the widespread belief that government offices, both local and national, are remote and inaccessible. Many governments worldwide believe that the Internet has the potential to reverse this trend. "E-governance" is now in use in many countries, both locally and nationally, allowing members of the public to write to candidates and participate in political discussions. This mechanism can help to create a dialogue between the authorities and their citizens, opening a channel for citizens to have a voice in shaping their society and so enhance their well-being.

The Internet makes us citizens of the global village and, for many people, this creates a feeling that they can make an impact on seemingly distant societies or groups. The phenomenon of online volunteering has been, for many surfers, a way of transferring this feeling to a reality. Online volunteers work in many different ways, including running online projects, translating important materials from one language to another, offering legal support, designing web sites which help populations in need, and creating study materials for online use (Amichai-Hamburger, 2008). One interesting example is a volunteering project managed by Elizabeth and Tim Rose from Canada (Rose and Rose, 2005). Their project is part of the Childcare and Adoption Society of Zambia, which aims to improve the life of homeless children in Zambia. Elizabeth and Tim, a mother and son who work from home, use the Internet to identify and contact potential donors of sports equipment. For those willing to contribute, they build a mechanism for collecting the donated items and shipping them to Zambia. (It is also important to note that Tim, an active partner in this project, is confined to a wheelchair.) Online volunteering affects the well-being of both the recipient and the donor, as Angie Ramaroson (2005), an online volunteer, explains: "Online volunteering is an opportunity to get involved in far-away communities with problems or issues close to your heart. It's a chance to change their world and your world as well."

Closing the digital gap

The Internet can only serve as a tool to enhance well-being in places where it is accessible. Therefore, it is very important that any discussion on the Internet and well-being include the topic of the digital gap – the fact that there are many poor countries where the Internet is not available to the majority of people (Norris, 2001). Even in the USA, there are many on lower incomes, particularly African Americans, who are not Internet users (Brodie *et al.*, 2000). One project which aims to combat the digital gap is the One Laptop per Child (OLPC) project. The project was launched in January 2005, in the MIT Media Lab, to develop a $100 laptop, a form of technology that could revolutionize how we educate the world's children. To achieve this goal, a new, nonprofit association, OLPC, independent of MIT, was created (http://wiki.laptop.org/go/One_Laptop_per_Child). It is hoped that by significantly reducing the cost of computers and Internet access, greater numbers of people around the world will be able to benefit from the Internet.

Addiction

Many personality characteristics have been linked to addiction, for example, introversion and sensation seeking (Shotton, 1991), tendency to boredom, self-consciousness, loneliness, social anxiety (Loytsker and Aiello, 1997), and poor self-esteem (Armstrong *et al.*, 2000). When assessing the relationship between Internet use and addiction, it is important to understand that a significant percentage of addictive surfers are addicts of a different sort offline and, in this case, have found another outlet for their proclivity.

Internet addiction is a very broad umbrella term. It is more useful to review specific Internet addictions, for example, addictions associated with chat, information, sex, and so on. To engage in depth with this problem and attempt to counter these behaviors, it is first necessary to gain a full understanding of the relationship between addiction to specific services and specific personality characteristics.

Enhancing Internet abilities

Future technology will result in the Internet having an even bigger impact on our lives. The capabilities of the Internet as an interactive tool that allows its users to design their own environment are growing. In the future, the Internet will be able to design an environment that suits the personality of the individual surfer (Amichai-Hamburger, 2002). This facility of the Internet to adapt its interactive capacity is related to the process of flow. Flow is an optimal experience that stems from people's perceptions of challenges and skills in given situations (Csikszentmihalyi, 1975: 50). Flow is a psychological state in which an individual feels cognitively efficient, motivated, and happy (Moneta and Csikszentmihalyi, 1996: 277). Basing their findings on self-reports of surfers, Chen *et al.* (2000) suggested that flow exists on the Internet, and that flow theory should be applied to Internet design so that Internet surfing will provide an enjoyable experience and positively affect surfer well-being. As net designers begin increasingly to work according to the principles of flow, the results for the user will become more significant.

Enhancing the virtual reality capabilities of the Internet will create an even more comprehensive, enjoyable environment that may impact the user in a variety of ways. One example may be the ability to deliver sophisticated therapy through the net. Virtual reality therapy (VRT) is a method of psychotherapy that uses virtual reality technology. It is currently used to treat patients with a variety of problems, such as burn pains (Hoffman *et al.*, 2000), fear of flying (Rothbaum *et al.*, 2002), and fear of spiders (Hoffman, 2004). VRT was shown to have long-term success rates (Choy *et al.*, 2007). As Internet technology becomes more sophisticated, VRT will be available for surfers on their Internet platforms.

Another advancement envisaged for Internet technology is one which will enable surfers around the world to communicate with one another in their own mother tongue, as each communication will be simultaneously translated. This, in itself, will enable more effective communication between different groups and different races (Amichai-Hamburger and McKenna, 2006). Other advancements in the realm of design would be the adaptation of web sites to meet the needs of particular groups; for example, web sites for the elderly could display fonts in a larger format that may be further enlarged, and options would be expressed more simply and clearly. Voice commands, rather than instructions through the keyboard, would also be a further advancement for this type of site.

It is, however, vital to understand that any and all attempts to enhance surfer well-being depend on the Internet being a secure, safe environment. It is of paramount importance that a balance be struck and maintained between freedom on the Internet and regulation. This balance is important for each specific element of the net; for example, a web site offering group therapy will not succeed if the participating surfer does not feel protected when he or she discloses information. Policymakers, academics, and web designers have a responsibility to work together to explore ways in which to enhance the positive impact of the Internet on well-being.

References

Adams, N., Stubbs, D., and Woods, V. (2005). Psychological barriers to Internet usage among older adults in the UK. *Medical Informatics and the Internet in Medicine*, 30, 3–17.

Alao, A., Soderberg, M., Pohl, E., and Lolaalao, A. (2006). Cybersuicide: Review of the role of the Internet on suicide. *CyberPsychology & Behavior*, 9, 489–493.

Allbon, E., and Williams, P. (2002). Nasties in the net: Children and censorship on the Web. *New Library World*, 103(1172/3), 30–38.

Allport, G. W. (1954). *The Nature of Prejudice*. Cambridge, MA: Addison-Wesley.

Allwes, D., and Popovich, M. L. (2007). Empowering patients and researchers through a common health information. *Studies in Health Technology and Informatics*, 127, 219–228.

Amichai-Hamburger, Y. (2002). Internet and personality. *Computers in Human Behavior*, 18, 1–10.

(2005). Personality and the Internet. In Y. Amichai-Hamburger (Ed.), *The Social Net: Human behavior in cyberspace* (pp. 27–55). New York: Oxford University Press.

(2008). Interacting via Internet: Theoretical and practical aspects. In A. Barak (Ed.), *Psychological Aspects of Cyberspace: Theory, research, applications* (pp. 207–229). Cambridge: Cambridge University Press.

Amichai-Hamburger, Y., and Ben-Artzi, E. (2003). Loneliness and Internet use. *Computers in Human Behavior*, 19, 71–80.

Amichai-Hamburger, Y., and Furnham, A. (2007). The positive net. *Computers in Human Behavior*, 23, 1033–1045.

Amichai-Hamburger, Y., and McKenna, K. Y. A. (2006). The contact hypothesis reconsidered: Interacting via the Internet. *Journal of Computer-Mediated Communication*, 11(3), article 7. Retrieved April 15, 2008, from http://jcmc.indiana.edu/vol11/issue3/amichai-hamburger.html.

Amichai-Hamburger, Y., Wainapel, G., and Fox, S. (2002). On the Internet no one knows I am an introvert: Extroversion, neuroticism, and Internet interaction. *CyberPsychology & Behavior*, 5(2), 125–128.

Amir, Y. (1969). Contact hypothesis in ethnic relations. *Psychological Bulletin*, 71(5), 319–342.

(1976). The role of intergroup contact in change of prejudice and ethnic relations. In P. A. Katz (Ed.), *Towards the Elimination of Racism* (pp. 73–123). New York: Plenum Press.

Anderberg, P. (2007). Peer assistance for personal assistance: Analysis of online discussions about personal assistance from a Swedish Web forum for disabled people. *Disability & Society*, 22, 251–265.

Andersson, G. (2006). Internet-based cognitive-behavioral self-help for depression. *Expert Review of Neurotherapeutics*, 6, 1637–1642.

Anthony, K. (2006). Electronically delivered therapies. In C. Feltham and I. Horton (Eds.), *The SAGE Handbook of Counselling and Psychotherapy* (pp. 518–523). London: Sage.

Argyle, M. (1987). *The Psychology of Happiness*. London: Methuen.

Argyle, M., and Furnham, A. (1983). Sources of satisfaction and conflict in long-term relationships. *Journal of Marriage and the Family*, 45, 481–493.

Armstrong, L., Phillips, J. G., and Saling, L. L. (2000). Potential determinants of heavier Internet usage. *International Journal of Human Computer Studies*, 53, 537–550.

Ayers, S. L., and Kronenfeld, J. J. (2007). Chronic illness and health-seeking information on the Internet. *Health*, 11, 327–347.

Baker, A. (2002). What makes an online relationship successful? Clues from couples who met in cyberspace. *CyberPsychology & Behavior*, 5, 363–375.

(2008). Down the rabbit hole: The role of place in the initiation and development of online relationships. In A. Barak (Ed.), *Psychological Aspects of Cyberspace: Theory, research, applications* (pp. 163–184). Cambridge: Cambridge University Press.

Baker, A. J. (2005). *Double Click: Romance and commitment among online couples*. Cresskill, NJ: Hampton Press.

Baker, L., Wagner, T. H., Bundorf, M. K., and Singer, S. (2003). Use of the Internet for health information and communication: Reply. *Journal of the American Medical Association*, 290, 2257–2258.

Barak, A. (2004). Internet counseling. In C. E. Spielberger (Ed.), *Encyclopedia of Applied Psychology* (pp. 369–378). San Diego, CA: Academic Press.

(2005). Research and experiences in interventions in online groups: Opportunities and challenges aside: Difficulties and risks. *Mikbatz*, 10(2), 33–56. (Published in Hebrew.)

(2007a). Emotional support and suicide prevention through the Internet: A field project report. *Computers in Human Behavior*, 23, 971–984.

(2007b). Phantom emotions: Psychological determinants of emotional experiences on the Internet. In A. Joinson, K. McKenna, T. Postmes, and U. Reips (Eds.), *The Oxford Handbook of Internet Psychology* (pp. 303–330). Oxford: Oxford University Press.

Barak, A., Boniel-Nissim, M., and Suler, J. (2008). Fostering empowerment in online support groups. *Computers in Human Behavior*, 24, 1867–1883.

Barak, A., and Dolev-Cohen, M. (2006). Does activity level in online support groups for distressed adolescents determine emotional relief? *Counselling and Psychotherapy Research*, 6, 186–190.

Barak, A., and Gluck-Ofri, O. (2007). Degree and reciprocity of self-disclosure in online forums. *CyberPsychology & Behavior*, 10, 407–417.

Barak, A., and Hen, L. (2008). Exposure in cyberspace as means of enhancing psychological assessment. In A. Barak (Ed.), *Psychological Aspects of Cyberspace: Theory, research, applications* (pp. 129–162). Cambridge: Cambridge University Press.

Barak, A., Hen, L., Boniel-Nissim, M., and Shapira, N. (2008). A comprehensive review and a meta-analysis of the effectiveness of Internet-based psychotherapeutic interventions. *Journal of Technology in Human Services*, 26, 109–160.

Barak, A., and King, S. A. (2000). Editorial: The two faces of the Internet: Introduction to the Special Issue on the Internet and sexuality. *CyberPsychology & Behavior*, 3, 517–520.

Barak, A., and Wander-Schwartz, M. (2000). Empirical evaluation of brief group therapy conducted in an Internet chat room. *Journal of Virtual Environments*, 5, 1. Retrieved March 25, 2007, from www.brandeis.edu/pubs/jove/HTML/v5/cherapy3.htm.

Bargh, J. A., Fitzsimons, G. M., and McKenna, K. Y. A. (2003). The self, online. In S. J. Spencer, S. Fein, M. P. Zanna, and J. M. Olson (Eds.), *Motivated Social Perception* (pp. 195–213). Mahwah, NJ: Erlbaum.

Bargh, J. A., and McKenna, K. Y. A. (2004). The Internet and social life. *Annual Review of Psychology*, 55, 573–590.

Bargh, J. A., McKenna, K. Y. A., and Fitzsimons, G. M. (2002). Can you see the real me? Activation and expression of the "true self" on the Internet. *Journal of Social Issues*, 58, 33–48.

Baumeister, R. F., and Leary, M. R. (1995). The need to belong: Desire for interpersonal attachment as a fundamental human motivation. *Psychological Bulletin*, 117, 497–529.

Becker, K., Mayer, M., Nagenborg, M., El-Faddagh, M., and Schmidt, M. H. (2004). Parasuicide online: Can suicide websites trigger suicidal behavior in predisposed adolescents? *Nordic Journal of Psychiatry*, 58, 111–114.

Bell, B. S., and Kozlowski, S. W. J. (2002). A typology of virtual teams: Implications for effective leadership. *Group and Organization Management*, 27(1), 14–49.

Bellafiore, D. R., Colòn, Y., and Rosenberg, P. (2004). Online counseling groups. In R. Kraus, J. Zack, and G. Stricker (Eds.), *Online Counseling: A handbook for mental health professionals* (pp. 197–216). San Diego, CA: Elsevier Academic Press.

Ben-Ze'ev, A. (2005). Detattachment: The unique nature of online romantic relationships. In Y. Amichai-Hamburger (Ed.), *The Social Net: The social psychology of the Internet* (pp. 115–138). Oxford: Oxford University Press.

Bettencourt, B. A., and Sheldon, K. (2001). Social roles as a vehicle for psychological need satisfaction within groups. *Journal of Personality and Social Psychology*, 81, 1131–1143.

Biddle, B. J. (1979). *Role Theory: Expectations, identities and behaviors.* New York: Academic Press.

Blank, T. O., and Adams-Blodnieks, M. (2007). The who and the what of usage of two cancer online communities. *Computers in Human Behavior*, 23, 1249–1257.

Blit-Cohen, E., and Litwin, H. (2004). Elder participation in Cyberspace: A qualitative analysis of Israeli retirees. *Journal of Aging Studies*, 18, 385–398.

Boies, S. C., Cooper, A., and Osborne, C. S. (2004). Variations in Internet-related problems and psychosocial functioning in online sexual activities: Implications for social and sexual development of young adults. *CyberPsychology & Behavior*, 7, 207–230.

Borzekowski, D. L. G., and Rickert, V. I. (2001). Adolescents, the Internet and health: Issues of access and content. *Journal of Applied Developmental Psychology*, 22, 49–59.

Bowker, N., and Tuffin, K. (2002). Disability discourses for online identities. *Disability & Society*, 17, 327–344.

Boyd, D. M., and Ellison, N. B. (2008). Social network sites: Definition, history, and scholarship. *Journal of Computer-Mediated Communication*, 13, 210–230.

Branden, N. (1969). *The Psychology of Self-esteem*. New York: Bantam.

Brandon, D. P., and Hillingshead, A. B. (2007). Characterizing online groups. In A. Joinson, K. McKenna, T. Postmes, and U. Reips (Eds.), *The Oxford Handbook of Internet Psychology* (pp. 105–120). Oxford: Oxford University Press.

Brewer, M. B. (1991). The social self: On being the same and different at the same time. *Personality and Social Psychology Bulletin*, 17, 475–482.

Brewer, M. B., and Gardner, W. L. (1996). Who is this "we"? Levels of collective identity and self-representations. *Journal of Personality and Social Psychology*, 71, 83–93.

Brodie, M., Flournoy, R. E., Altman, D. E., Blendon, R. J., Benson, J. M., and Rosenbaum, M. D. (2000). Health information, the Internet, and the digital divide. *Health Affairs*, 19(6), 255–265.

Brown, R. J. (1978). Divided we fall: An analysis of relations between sections of a factory work force. In H. Tajfel (Ed.), *Differentiation Between Social Groups: Studies in the social psychology of intergroup relations* (pp. 395–429). London: Academic Press.

Bryant, J. A., Sanders-Jackson, A., and Smallwood, A. M. K. (2006). IMing, text messaging, and adolescent social networks. *Journal of Computer-Mediated Communication*, 11, 577–592.

Buchanan, H., and Coulson, N. S. (2007). Accessing dental anxiety online support groups: An exploratory qualitative study of motives and experiences. *Patient Education and Counseling*, 66, 263–269.

Bundorf, M. K., Wagner, T. H., Singer, S. J., and Baker, L. C. (2006). Who searches the Internet for health information? *Health Services Research*, 41, 819–836.

Campbell, H. (2004). Challenges created by online religious networks abstract. *Journal of Media and Religion*, 3(2), 81–99.

Carlbring, P., and Andersson, G. (2006). Internet and psychological treatment: How well can they be combined? *Computers in Human Behavior*, 22, 545–553.

Carpenter, B. D., and Buday, S. (2007). Computer use among older adults in a naturally occurring retirement community. *Computers in Human Behavior*, 23, 3012–3024.

Celio, A. A., Winzelberg, A. J., Wilfley, D. E., Eppstein-Herald, D., Springer, E. A., Dev, P., and Taylor, C. B. (2000). Reducing risk factors for eating disorders: Comparison of an Internet and a classroom-delivered psychoeducational program. *Journal of Consulting and Clinical Psychology*, 68, 650–657.

Chan, D. K.-S., and Cheng, G. H.-L. (2004). A comparison of offline and online friendship qualities at different stages of relationship development. *Journal of Social and Personal Relationships*, 21, 305–320.

Chang, T., Yeh, C. J., and Krumboltz, J. D. (2001). Process and outcome evaluation of an on-line support group for Asian American male college students. *Journal of Counseling Psychology*, 48, 319–329.

Charles, C., Gafni, A., and Whelan, T. (1997). Shared decision-making in the medical encounter: What does it mean? (Or it takes at least two to tango). *Social Science and Medicine*, 44, 681–692.

Chen, H. A., Wigand, R. T., and Nilan, M. (2000). Exploring web users' optimal flow experiences. *Information Technology and People*, 13, 263–281.

Cheng, G. H. L., Chan, D. K. S., and Tong, P. Y. (2006). Qualities of online friendships with different gender compositions and durations. *CyberPsychology & Behavior*, 9, 14–21.

Chiou, W. (2006). Adolescents' sexual self-disclosure on the Internet: Deindividuation and impression management. *Adolescence*, 41, 547–561.

Cho, S.-H. (2007). Effects of motivations and gender on adolescents' self-disclosure in online chatting. *CyberPsychology & Behavior*, 10, 339–345.

Choy, Y., Fyer, A. J., and Lipsitz, J. D. (2007). Treatment of specific phobia in adults. *Clinical Psychology Review*, 27, 266–286.

Cialdini, R. B. (1984). *The Psychology of Persuasion*. New York: Quill William Morrow.

Cole, J. (2001). *The UCLA Internet Report: Surveying the digital future, year 2*. Los Angeles: UCLA Center for Communication Policy. Retrieved February 3, 2008, from www.digitalcenter.org/pdf/InternetReportYearTwo.pdf.

Colòn, Y., and Friedman, B. (2003). Conducting group therapy online. In S. Goss and K. Anthony (Eds.), *Technology in Counselling and Psychotherapy: A practitioner's guide* (pp. 59–74). Houndmills, UK: Palgrave Macmillan.

Cook, S. W. (1962). The systematic analysis of socially significant events: A strategy for social research. *Journal of Social Issues*, 18(2), 66–84.

Covey, S. (1989). *The Seven Habits of Highly Effective People*. New York: Simon & Schuster.

Csikszentmihalyi, M. (1975). *Beyond Boredom and Anxiety*. San Francisco: Jossey-Bass.

Cummings, J. N., Sproull, L., and Kiesler, S. B. (2002). Beyond hearing: Where the real-world and online support meet. *Group Dynamics*, 6, 78–88.

Davison, K. P., Pennebaker, J. W., and Dickerson, S. S. (2000). Who talks? The social psychology of illness support groups. *The American Psychologist*, 55, 205–217.

Dietz-Uhler, B., Bishop-Clark, C., and Howard, E. (2005). Formation of and adherence to a self-disclosure norm in an online chat. *CyberPsychology & Behavior*, 8, 114–120.

Dombrowski, S. C., Gischlar, K. L., and Durst, T. (2007). Safeguarding young people from cyber pornography and cyber sexual predation: A major dilemma of the Internet. *Child Abuse Review*, 16, 153–170.

Dubow, E. F., Lovoko, K. R., and Kausch, D. F. (1990). Demographic differences in adolescents' health concerns and perceptions of helping agents. *Journal of Clinical Child Psychology*, 19, 44–54.

Dunning, T. (2004). Aging, activities, and the Internet. *Adaptation and Aging*, 28, 75–76.

Durkin, K. F. (2004). The Internet as a milieu for the management of a stigmatized sexual identity. In D. D. Waskul (Ed.), *Net.SeXXX: Readings on sex, pornography, and the Internet* (pp. 131–147). New York: Peter Lang.

Eastman, J. K., and Iyer, R. (2004). The elderly's uses and attitudes towards the Internet. *Journal of Consumer Marketing*, 21(3), 208–220.

Ellerman, E. (2007). The Internet in context. In J. Gackenbach (Ed.), *Psychology and the Internet: Intrapersonal, interpersonal and transpersonal implications*, 2nd edn. (pp. 11–36). San Diego, CA: Elsevier Academic Press.

Ellison, N. B., Steinfield, C., and Lampe, C. (2007). The benefits of Facebook "Friends": Social capital and college students' use of online social network sites. *Journal of Computer-Mediated Communication*, 12, 1143–1168.

Epstein, J., and Klinkenberg, W. D. (2001). From Eliza to Internet: A brief history of computerized assessment. *Computers in Human Behavior*, 17, 295–314.

Epstein, Y. M., Rosenberg, H. S., Grant, T. V., and Hemenway, N. (2002). Use of the Internet as the only outlet for talking about infertility. *Fertility and Sterility*, 78, 507–514.

Erikson, E. H. (1968). *Identity, Youth and Crisis*. New York: Norton.

Etter, J. (2006). The Internet and the industrial revolution in smoking cessation counselling. *Drug and Alcohol Review*, 25, 79–84.

Eysenck, H. J., and Eysenck, S. B. G. (1975). *Manual of the Eysenck Personality Questionnaire (adult and junior)*. London: Hodder & Stoughton.

Farnham, S., Cheng, L., Stone, L., Zaner-Godsey, M., Hibbeln, C., Syrjala, K., Clark, A. M., and Abrams, J. (2002). *HutchWorld: Clinical study of computer-mediated social support for cancer patients and their caregivers. Conference on Human Factors in Computing Systems*. Proceedings of the SIGCHI Conference on Human Factors in Computing Systems (pp. 375–382).

Feldman, M. D. (2000). Munchausen by Internet: Detecting factitious illness and crisis on the Internet. *Southern Medical Journal*, 93, 669–672.

Fenichel, M., Suler, J., Barak, A., Zelvin, E., Jones, G., Munro, K., Meunier, V., and Walker-Schmucker, W. (2002). Myths and realities of online clinical work. *CyberPsychology & Behavior*, 5, 481–497.

Feshbach, S., and Weiner, B. (1991). *Personality*. Lexington, MA: D. C. Heath.

Finfgeld, D. L. (1999). Psychotherapy in cyberspace. *Journal of the American Psychiatric Nurses Association*, 5(4), 105–110.

Fleming, M. J., Greentree, S., Cocotti-Muller, D., Elias, K. A., and Morrison, S. (2006). Safety in cyberspace: Adolescents' safety and exposure online. *Youth & Society*, 38, 135–154.

Flynn, K. E., Smith, M. A., and Freese, J. (2006). When do older adults turn to the Internet for health information? Findings from the Wisconsin longitudinal study. *Journal of Global Information Management*, 21, 1–7.

Fox, S., Rainie, L., Larsen, E., Horrigan, J., Lenhart, A., Spooner, T., and Carter, C. (2001). Wired seniors. *The Pew Internet and American Life Project*. Retrieved March 22, 2007, from www.pewinternet.org/Reports/2006/Are-Wired-Seniors-Sitting-Ducks.aspx.

Fozard, J. L., and Kearns, W. D. (2007). Technology, aging, and communication. In G. Lesnoff-Caravaglia (Ed.), *Gerontechnology: Growing old in a technological society* (pp. 271–291). Springfield, IL: Charles C. Thomas.

Fuglsang, L. (2005). IT and senior citizens: Using the Internet for empowering active citizenship. *Science, Technology, & Human Values*, 30, 468–495.

Galegher, J., and Kraut, R. E. (1994). Computer-mediated communication for intellectual teamwork: An experiment in group writing. *Information Systems Research*, 5(2), 110–138.

Gerber, B. S., and Eiser, A. R. (2001). The patient–physician relationship in the Internet age: Future prospects and the research agenda. *Journal of Medical and Internet Research*, 3, article 15.

Goldman, D. (2003). Sex, lies and the Internet. *Yediot*, 12, Oct. 1. (Published in Hebrew.)

Golkaramnay, V., Bauer, S., Haug, S., Wolf, M., and Kordy, H. (2007). The exploration of the effectiveness of group therapy through an Internet chat as aftercare: A controlled naturalistic study. *Psychotherapy and Psychosomatics*, 76, 219–225.

Gray, N. J., and Klein, J. D. (2006). Adolescents and the Internet: Health and sexuality information. *Current Opinion in Obstetrics and Gynecology*, 18, 519–524.

Gray, N. J., Klein, J. D., Noyce, P. R., Sesselberg, T. S., and Cantrill, J. A. (2005). Health information-seeking behaviour in adolescence: The place of the Internet. *Social Science & Medicine*, 60, 1467–1478.

Grohol, J. M. (2004). Online counseling: A historical perspective. In R. Kraus, J. Zack, and G. Stricker (Eds.), *Online Counseling: A handbook for mental health professionals* (pp. 51–68). San Diego, CA: Elsevier Academic Press.

Gross, E. F. (2004). Adolescent Internet use: What we expect, what teens report. *Applied Developmental Psychology*, 25, 633–649.

Grossman, S. L. (2007). Faithful build a Second Life for religion online. *USA Today*. Retrieved February 15, 2008, from www.usatoday.com/tech/gaming/2007-04-01-second-life-religion_N.htm.

Hamburger, Y. (1994). The contact hypothesis. *Journal of Basic and Applied Social Psychology*, 15, 339–358.

Hamburger, Y. A., and Ben-Artzi, E. (2000). The relationship between extraversion and neuroticism and the different uses of the Internet. *Computers in Human Behavior*, 16, 441–449.

Hardey, M. (2002). Life beyond the screen: Embodiment and identity through the Internet. *The Sociological Review*, 50(4), 570–585.

Hardt, J. H., and Hollis-Sawyer, L. (2007). Older adults seeking healthcare information on the Internet. *Educational Gerontology*, 33, 561–572.

Haythornthwaite, C. (2007). Social networks and online community. In A. Joinson, K. McKenna, T. Postmes, and U. Reips (Eds.), *The Oxford Handbook of Internet Psychology* (pp. 121–138). Oxford: Oxford University Press.

Haythornthwaite, C., and Hagar, C. (2004). The social worlds of the Web. *Annual Review of Information Science and Technology*, 39, 311–346.

Heinicke, B. E., Paxton, S. J., McLean, S. A., and Wertheim, E. H. (2007). Internet-delivered targeted group intervention for body dissatisfaction and disordered eating in adolescent girls: A randomized controlled trial. *Journal of Abnormal Child Psychology*, 35, 379–391.

Henderson, S., and Gilding, M. (2004). "I've never clicked this much with anyone in my life": Trust and hyperpersonal communication in online friendships. *New Media and Society*, 6, 487–506.

Hlebec, V., Manfreda, K. L., and Vehovar, V. (2006). The social support networks of Internet users. *New Media & Society*, 8, 9–32.

Hoffman, H. G. (2004). Virtual-reality therapy. *Scientific American*, 291, 58–65.

Hoffman, H. G., Doctor, J. N., Patterson, D. R., Carrougher, G. J., and Furness, T. A., III. (2000). Use of virtual reality for adjunctive treatment of adolescent burn pain during wound care: A case report. *Pain*, 85, 305–309.

Hoover, S. M., Clark, L. S., and Rainie, L. (2004). Faith Online. *Pew Internet & American Life*, Washington, DC.

Hopps, S. L., Pépin, M., and Boisvert, J. M. (2003). The effectiveness of cognitive-behavioral group therapy for loneliness via Internet-relay-chat among people with physical disabilities. *Psychotherapy: Theory/Research/Practice/Training*, 40(1–2), 136–147.

Houston, G. T. K., Cooper, L. A., and Ford, D. E. (2002). Internet support groups for depression: A 1-year prospective cohort study. *American Journal of Psychiatry*, 159, 2062–2068.

Hu, Y., Wood, F. J., Smith, V., and Westbrook, N. (2004). Friendships through IM: Examining the relationship between instant messaging and intimacy. *Journal of Computer-Mediated Communication*, 10(1). Retrieved March 22, 2008, from http://jcmc.indiana.edu/vol10/issue1/hu.html.

Jerin, R. A., and Dolinsky, B. (2007). Cyber-victimisation and online dating. In M. T. Whitty, A. J. Baker, and J. A. Inman (Eds.), *Online Matchmaking* (pp. 147–158). London: Palgrave Macmillan.

Joinson, A. N. (2007). Disinhibition and the Internet. In J. Gackenbach (Ed.), *Psychology and the Internet: Intrapersonal, interpersonal and transpersonal implications*, 2nd edn. (pp. 76–92). San Diego, CA: Elsevier Academic Press.

Joinson, A. N., and Paine, C. B. (2007). Self-disclosure, privacy and the Internet. In A. Joinson, K. McKenna, T. Postmes, and U. Reips (Eds.), *The Oxford Handbook of Internet Psychology* (pp. 237–252). Oxford: Oxford University Press.

Jourard, S. M. (1971). *Self-disclosure: An experimental analysis of the transparent self.* New York: Wiley.

Jucks, R., and Bromme, R. (2007). Choice of words in doctor–patient communication: An analysis of health-related Internet sites. *Health Communication*, 21, 267–277.

Karavidas, M., Lim, N. K., and Katsikas, S. L. (2005). The effects of computers on older adult users. *Computers in Human Behavior*, 21, 679–711.

Kraus, R. (2004). Ethical and legal considerations for providers of mental health services online. In R. Kraus, J. Zack, and G. Stricker (Eds.), *Online Counseling: A handbook for mental health professionals* (pp. 123–144). San Diego, CA: Elsevier Academic Press.

Kraut, R., Keisler, S., Boneva, B., Cummings, J., Helgeson, V., and Crawford, A. (2002). Internet paradox revisited. *Journal of Social Issues*, 58, 49–74.

Kraut, R., Lundmark, V., Patterson, M., Kiesler, S., Mukopadhyay, T., and Scherlis, W. (1998). Internet paradox: A social technology that reduces social involvement and psychological well-being. *American Psychologist*, 53, 1017–1031.

Lamberg, L. (1997). Online support group helps patients live with, learn more about the rare skin cancer CTCL-MF. *Journal of the American Medical Association*, 277, 1422–1423.

Larson, R. (1978). Thirty years of research on the subjective well-being of older Americans. *Journal of Gerontology*, 33, 109–125.

Lawson, H. M., and Leck, K. (2006). Dynamics of Internet dating. *Social Science Computer Review*, 24, 189–208.

Ley, B. L. (2007). Vive Les Roses! The architecture of commitment in an online pregnancy and mothering group. *Journal of Computer-Mediated Communication*, 12, 1388–1408.

Lieberman, M. A., and Goldstein, B. A. (2006). Not all negative emotions are equal: The role of emotional expression in online support groups for women with breast cancer. *Psycho-Oncology*, 15, 160–168.

Link, B. G., Struening, E. L., Neese-Todd, S., Asmussen, S., and Phelan, J. C. (2001). Stigma as a barrier to recovery: The consequences of stigma for the self-esteem of people with mental illnesses. *Psychiatric Services*, 52, 1621–1626.

Linville, P. W. (1987). Self-complexity as a cognitive buffer against stress-related illness and depression. *Journal of Personality and Social Psychology*, 52, 663–676.

Livingstone, S., and Bober, M. (2005). *UK Children Go Online: Final report to the Economic and Social Research Council*. London: Economic and Social Research Council.

Loytsker, J., and Aiello, J. R. (1997). Internet addiction and its personality correlates. Poster presented at the annual meeting of the Eastern Psychological Association, Washington, DC, USA, April.

Maczewski, M. (2002). Exploring identities through the Internet: Youth experiences online. *Child & Youth Care Forum*, 31, 111–129.

Madell, D. E., and Muncer, S. J. (2007). Control over social interactions: An important reason for young people's use of the Internet and mobile phones for communication. *CyberPsychology & Behavior*, 10, 137–140.

Mallen, M. J., Vogel, D. L., Rochlen, A. B., and Day, S. X. (2005). Online counseling: Reviewing the literature from a counseling psychology framework. *The Counseling Psychologist*, 33, 819–871.

Manhal-Baugus, M. (2001). E-therapy: Practical, ethical, and legal issues. *CyberPsychology & Behavior*, 4, 551–563.

Marcus, B., Machilek, F., and Schütz, A. (2006). Personality in cyberspace: Personal Web sites as media for personality expressions and impressions. *Journal of Personality and Social Psychology*, 90, 1014–1031.

Marziali, E. (2006). Developing evidence for an Internet-based psychotherapeutic group intervention. *Journal of Evidence-Based Social Work*, 3, 149–165.

Maslow, A. (1971). *The Farther Reaches of Human Nature*. New York: The Viking Press.

McKenna, K. Y. A. (2001). Demarginalizing the sexual self. *Journal of Sex Research*, 38, 302–312.

(2007a). Through the Internet looking glass: Expressing and validating the true self. In A. Joinson, K. McKenna, T. Postmes, and U. Reips (Eds.),

The Oxford Handbook of Internet Psychology (pp. 205–222). Oxford: Oxford University Press.

(2007b). A progressive affair: Online dating to real world mating. In M. T. Whitty, A. J. Baker, and J. A. Inman (Eds.), *Online Matchmaking* (pp. 112–124). London: Palgrave Macmillan.

McKenna, K., and Bargh, J. (1998). Coming out in the age of the Internet identity: Demarginalization through virtual group participation. *Journal of Personality and Social Psychology*, 75, 681–694.

McKenna, K. Y. A., and Bargh, J. A. (2000). Plan 9 from cyberspace: The implications of the Internet for personality and social psychology. *Personality and Social Psychology Review*, 4(1), 57–75.

McKenna, K. Y. A., Green, A. S., and Gleason, M. J. (2002). Relationship formation on the Internet: What's the big attraction? *Journal of Social Issues*, 58, 9–32.

McKenna, K., and Seidman, G. (2005). You, me, and we: Interpersonal processes in electronic groups. In Y. Amichai-Hamburger (Ed.), *The Social Net: Human behavior in cyberspace* (pp. 191–217). New York: Oxford University Press.

McKenna, K. Y. A., and West, K. J. (2007). Give me that online-time religion: The role of the Internet in spiritual life. *Computers in Human Behavior*, 23, 942–954.

McMellon, C. A., and Schiffman, L. G. (2002). Cybersenior empowerment: How some older individuals are taking control of their lives. *Journal of Applied Gerontology*, 21(2), 157–175.

McMullan, M. (2006). Patients using the Internet to obtain health information: How this affects the patient–health professional relationship. *Patient Education and Counseling*, 63, 24–28.

Meischke, H., Eisenberg, M., Rowe, S., and Cagle, A. (2005). Do older adults use the Internet for information on heart attacks? Results from a survey of seniors in King County, WA. *Heart and Lung: The journal of acute and critical care*, 34, 3–12.

Mileham, B. L. A. (2003). Online infidelity in Internet chat rooms. Unpublished doctoral dissertation, University of Florida, Gainesville, FL, USA.

Mitchell, K. J., Finkelhor, D., and Wolak, J. (2001). Risk factors and impact of online sexual solicitation of youth. *Journal of the American Medical Association*, 285(23), 1–4.

(2003). The exposure of youth to unwanted sexual material on the Internet: A national survey of risk, impact and prevention. *Youth & Society*, 34(3), 330–358.

(2007). Online requests for sexual pictures from youth: Risk factors and incident characteristics. *The Journal of Adolescent Health*, 41, 196–203.

Moneta, G. B., and Csikszentmihalyi, M. (1996). The effect of perceived challenges and skills on the quality of subjective experience. *Journal of Personality*, 64, 275–310.

Morahan-Martin, J. (2000). Women and the Internet: Promise and perils. *CyberPsychology & Behavior*, 3, 683–691.

(2004). Paradoxes in the impact of the Internet on women. In K. Morgan, C. A. Brebbia, J. Sanchez, and A. Voiskounsky (Eds.), *Human Perspectives in the Internet Society: Culture, psychology and gender* (pp. 275–285). Southampton, UK: WIT Press.

Myers, D. G. (1992). *The Pursuit of Happiness*. New York: Quill.

Nettle, D. (2005). *Happiness: The science behind your smile*. New York: Oxford University Press.

Norris, P. (2001). *Digital Divide? Civic engagement, information poverty, and the Internet worldwide*. Cambridge: Cambridge University Press.

O'Hara, K. (2004). Curb cuts on the information highway: Older adults and the Internet. *Technical Communication Quarterly*, 13(4), 423–445.

Owen, J. E., Klapow, J. C., Roth, D. L., Nabell, L., and Tucker, D. C. (2004). Improving the effectiveness of adjuvant psychological treatment for women with breast cancer: The feasibility of providing online support. *Psycho-Oncology*, 13, 281–292.

Owen, J. E., Klapow, J. C., Roth, D. L., Shuster, J. L., and Bellis, J. (2005). Randomized pilot of a self-guided Internet coping group for women with early-stage breast cancer. *Annals of Behavioral Medicine*, 30, 54–64.

Parks, M. R., and Floyd, K. (1996). Making friends in cyberspace. *Journal of Communication*, 46, 80–97.

Podolsky, D. K. (1998). Patients, gastroenterologists, and the World Wide Web. *Gastroenterology*, Jan., 114(1), 5.

Preston, J. M. (2007). From mediated environments to the development of consciousness. In J. Gackenbach (Ed.), *Psychology and the Internet: Intrapersonal, interpersonal and transpersonal implications*, 2nd edn. (pp. 277–308). San Diego, CA: Elsevier Academic Press.

Qian, H., and Scott, C. R. (2007). Anonymity and self-disclosure on Weblogs. *Journal of Computer-Mediated Communication*, 12, 1428–1451.

Ramaroson, H. A. (2005). *Online Volunteering Services: Stories*. Retrieved July 27, 2007, from www.onlinevolunteering.org/en/org/stories/online_volunteers_of_2005.html.

Rau, P. P., and Hsu, J. (2005). Interaction devices and Web design for novice older users. *Educational Gerontology*, 31, 19–40.

Rheingold, H. (2000). *The Virtual Community: Homesteading on the electronic frontier*, rev. edn. Cambridge, MA: MIT Press.

Rideout, V. J., Vandewater, E. A., and Wartella, E. A. (2003). *Zero to Six: Electronic media in the lives of infants, toddlers, and preschoolers*. Menlo Park, CA: The Henry J. Kaiser Family Foundation.

Ritterband, L. M., Gonder-Frederick, L. A., Cox, D. J., Clifton, A. D., West, R. W., and Borowitz, S. M. (2003). Internet interventions: In review, in use, and into the future. *Professional Psychology: Research and practice*, 34, 527–534.

Riva, G., Gaggioli, A., Villani, D., Preziosa, A., Morganti, F., Corsi, R., Faletti, G., and Vezzadini, L. (2007). NeuroVR: An open source virtual reality platform for clinical psychology and behavioral neurosciences. *Studies in Health Technology and Informatics*, 125, 394–399.

Rochlen, A. B., Land, L. N., and Wong, Y. J. (2004). Male restrictive emotionality and evaluations of online versus face-to-face counseling. *Psychology of Men and Masculinity*, 5, 190–200.

Rogers, C. (1961). *On Becoming a Person: A therapist's view of psychotherapy*. London: Constable.

Rose, E., and Rose, T. (2005). *Online Volunteering Services: Stories*. Retrieved July 27, 2007, from www.onlinevolunteering.org/en/org/stories/online_volunteers_of_2005.html#3.

Rothbaum, B. O., Hodges, L., Anderson, P. L., Price, L., and Smith, S. (2002). Twelve-month follow-up of virtual reality and standard exposure therapies for the fear of flying. *Journal of Consulting and Clinical Psychology*, 70, 428–432.

Rotondi, A. J., Sinkule, J., and Spring, M. (2005). An interactive Web-based intervention for persons with TBI and their families: Use and evaluation by female significant others. *Journal of Head Trauma Rehabilitation*, 20, 173–185.

Rubin, Z. (1975). Disclosing oneself to a stranger: Reciprocity and its limits. *Journal of Experimental Social Psychology*, 11, 233–260.

Safran, C. (2003). Collaborative edge: Patient empowerment for vulnerable populations. *International Journal of Medical Informatics*, 69, 185–190.

Schottenbauer, M. A., Rodriguez, B. F., Glass, C. R., and Arnkoff, D. B. (2004). Computers, anxiety, and gender: An analysis of reactions to the Y2K computer problem. *Computers in Human Behavior*, 20(1), 67–83.

Senge, P. (1990). *The Fifth Discipline: The art and practice of the learning organization*. New York: Doubleday.

Shapira, N., Barak, A., and Gal, I. (2007). Promoting older adults' well-being through Internet training and use. *Aging & Mental Health*, 11, 477–484.

Shotton, M. (1991). The costs and benefits of computer addiction. *Behaviour and Information Technology*, 10, 219–230.

Skinner, A., and Zack, J. S. (2004). Counseling and the Internet. *American Behavioral Scientist*, 48, 434–446.

Stefanone, M. A., and Jang, C.-Y. (2008). Writing for friends and family: The interpersonal nature of blogs. *Journal of Computer-Mediated Communication*, 13, 123–140.

Stephan, W. G., and Stephan, C. W. (1996). Cognition and affect in stereotyping: Parallel interactive networks. In D. W. Mackie and D. L. Hamilton (Eds.), *Affect, Cognition and Stereotyping: Interactive processes in group perception* (pp. 111–136). Orlando, FL: Academic Press.

(2001). *Improving Intergroup Relations*. Newbury, CA: Sage.

Strecher, V. (2007). Internet methods for delivering behavioral and health-related interventions (eHealth). *Annual Review of Clinical Psychology*, 3, 53–76.

Suler, J. R. (2002). Identity management in cyberspace. *Journal of Applied Psychoanalytic Studies*, 4, 455–459.

(2004a). The online disinhibition effect. *CyberPsychology & Behavior*, 7, 321–326.

(2004b). The psychology of text relationships. In R. Kraus, J. Zack, and G. Stricker (Eds.), *Online Counseling: A handbook for mental health professionals* (pp. 19–50). San Diego, CA: Elsevier Academic Press.

(2008). Cybertherapeutic theory and techniques. In A. Barak (Ed.), *Psychological Aspects of Cyberspace: Theory, research, applications* (pp. 102–128). Cambridge: Cambridge University Press.

Tajfel, H. (1978). Social categorization, social identity and social comparison. In H. Tajfel (Ed.), *Differentiation Between Social Groups: Studies in the social psychology of intergroup relations* (pp. 61–76). London: Academic Press.

Tajfel, H., and Turner, J. C. (1986). The social identity theory of inter-group behavior. In S. Worchel and L. W. Austin (Eds.), *Psychology of Intergroup Relations* (pp. 7–24). Chicago: Nelson-Hall.

Tanis, M. (2007). Online social support groups. In A. Joinson, K. McKenna, T. Postmes, and U. Reips (Eds.), *The Oxford Handbook of Internet Psychology* (pp. 139–154). Oxford: Oxford University Press.

Thoits, P. (1986). Social support as coping assistance. *Journal of Consulting and Clinical Psychology*, 54, 416–423.

Thorndike, E. (1920). A constant error in psychological ratings. *Journal of Applied Psychology*, 4, 469–477.

Tidwell, L. C., and Walther, J. B. (2002). Computer-mediated communication effects on disclosure, impressions, and interpersonal evaluations: Getting to know one another a bit at a time. *Human Communication Research*, 28, 317–348.

Tsai, C. C., Lin, S. J., and Tsai, M. J. (2001). Developing the Internet Attitude Scale for high school students. *Computers and Education*, 37, 41–51.

Turkle, E. (2004). Whither psychoanalysis in computer culture? *Psychoanalytic Psychology*, 21, 16–30.

Turkle, S. (1995). *Life on the Screen: Identity in the age of the Internet.* New York: Simon & Schuster.

Tynes, B. M. (2007). Internet safety gone wild? Sacrificing the educational and psychosocial benefits of online social environments. *Journal of Adolescent Research*, 22, 575–584.

Uchino, B. N., Cacioppo, J. T., and Kiecolt-Glaser, J. K. (1996). The relationship between social support and physiological processes: A review with emphasis on underlying mechanisms and implications for health. *Psychological Bulletin*, 119, 488–531.

Valcke, M., Schellens, T., Van Keer, H., and Gerarts, M. (2007). Primary school children's safe and unsafe use of the Internet at home and at school: An exploratory study. *Computers in Human Behavior*, 23, 2838–2850.

Valkenburg, P. M., and Peter, J. (2007a). Preadolescents' and adolescents' online communication and their closeness to friends. *Developmental Psychology*, 43, 267–277.

(2007b). Online communication and adolescent well-being: Testing the stimulation versus the displacement hypothesis. *Journal of Computer-Mediated Communication*, 12, 1169–1182.

Valkenburg, P. M., Peter, J., and Schouten, A. P. (2006). Friend networking sites and their relationship to adolescents' well-being and social self-esteem. *CyberPsychology & Behavior*, 9, 584–590.

Valkenburg, P. M., and Soeters, K. E. (2001). Children's positive and negative experiences with the Internet. *Communication Research*, 28, 652–675.

van Uden-Kraan, C. F., Drossaert, C. H. C., Taal, E., Lebrun, C. E. I., Drossaers-Bakker, K. W., Smit, W. M., Seydel, E. R., and van de Laar, M. A. F. J. (2008). Coping with somatic illnesses in online support groups: Do the feared disadvantages actually occur? *Computers in Human Behavior*, 24, 309–324.

Velillas, J. J. L., Hernández-Aguilar, M.-T., Alonso, C. R. P., Díaz-Gómez, M., Papí, A. G., Espuelas, C. F., and Talayero, J. M. P. (2007). A breastfeeding e-learning project based on a web forum. *Breastfeeding Medicine*, 2, 219–228.

Walther, J. B. (1996). Computer-mediated communication: Impersonal, interpersonal, and hyperpersonal interaction. *Communication Research*, 23(1), 3–43.

Wasko, M. M., and Faraj, S. (2000). "It is what one does": Why people participate and help others in electronic communities of practice. *Journal of Strategic Information Systems*, 9(2–3), 155–173.

Watson, G., and Johnson, D. (1972). *Social Psychology: Issues and insights*. Philadelphia: J. B. Lippincott.

Weinstein, P. K. (2006). A review of weight loss programs delivered via the Internet. *The Journal of Cardiovascular Nursing*, 21, 251–258.

Wellman, B., Salaff, J., Dimitrova, D., Garton, L., Gulia, M., and Haythornthwaite, C. (1996). Computer networks as social networks: Collaborative work, telework, and virtual community. *Annual Review of Sociology*, 22, 213–238.

White, H., McConnell, E., Clipp, E., Branch, L. G., Sloane, R., Pieper, C., and Box, T. L. (2002). A randomized controlled trial of the psychosocial impact of providing Internet training and access to older adults. *Aging & Mental Health*, 6, 213–221.

Whitty, M. T., and Carr, A. N. (2006). *Cyberspace Romance: The psychology of online relationships*. London: Palgrave Macmillan.

Wilder, D. A. (1993). The role of anxiety in facilitating stereotypic judgments of out-group behavior. In D. M. Mackie and D. L. Hamilton (Eds.), *Affect, Cognition, and Stereotyping: Interactive processes in group perception* (pp. 87–109). San Diego, CA: Academic Press.

Wishart, J. M., Oades, C. E., and Morris, M. (2007). Using online role play to teach Internet safety awareness. *Computers & Education*, 48, 460–473.

Wolak, J., Mitchell, K., and Finkelhor, D. (2007). Unwanted and wanted exposure to online pornography in a national sample of youth Internet users. *Pediatrics*, 119, 247–257.

Wright, E. R., Gronfein, W. P., and Owens, T. J. (2000). Deinstitutionalization, social rejection, and the self-esteem of former mental patients. *Journal of Health and Social Behavior*, 41(1), 68–90.

Xie, B. (2007). Using the Internet for offline relationship formation. *Social Science Computer Review*, 25, 396–404.

Ybarra, M. L. (2004). Linkages between depressive symptomatology and Internet harassment among young regular Internet users. *CyberPsychology & Behavior*, 7, 247–257.

Ybarra, M. L., Eaton, W. W., and Bickman, L. (2005). Internet-based mental health interventions. *Mental Health Services Research*, 7, 75–87.

Ybarra, M. L., and Mitchell, K. J. (2007). Prevalence and frequency of Internet harassment instigation: Implications for adolescent health. *The Journal of Adolescent Health*, 41, 189–195.

Ybarra, M. L., and Suman, M. (2006). Help-seeking behavior and the Internet: A national survey. *International Journal of Medical Information*, 75(1), 29–41.

Young, K. S., Griffin-Shelley, E., Cooper, A., O'Mara, J., and Buchanan, J. (2000). Online infidelity: A new dimension in couple relationships with implications for evaluation and treatment. In A. Cooper (Ed.), *Cybersex: The dark side of the force* (pp. 59–74). Philadelphia: Brunner Routledge.

Zander, A., Stodland, E., and Wolfe, D. (1960). Unity of group, identification with group, and self-esteem of members. *Journal of Personality*, 28, 463–478.

3 Information, innovation, and society

Steven L. Goldman

That we are in the midst of an information "revolution" is a staple of media coverage of digital electronic technologies and their transformative social impact. This characterization seems so justified by our own personal and social experiences that it is easy to lose critical perspective on the subject of information. Is the relationship between information and society really any different today from what it had been prior to the late twentieth century? What *is* information that it is capable of causing social change? More broadly, what are technological innovations that they are able to cause social change and how has our well-being been affected by information technology-driven change? That is, are we better off, personally and socially, because of the digital information "revolution"? What follows explores answers to these questions.

Revolution or evolution?

When talk of revolution comes up it is worth keeping in mind the dictum of art historian Walter Friedlander (1957) that every so-called *revolution* becomes an *evolution* when its precursors are properly understood. While revolution implies a radical discontinuity with the past, evolution implies a deep continuity with the past, the episodic introduction of discontinuities into these continuities leading to the emergence of true novelties. History is far more central to the generation of evolutionary phenomena than to revolutionary ones. So, is society today experiencing a revolutionary change in the historical relationship between information and society, or an evolutionary change in that relationship? If the latter, our understanding of the social impact of digital information technologies, and our ability to assess their effect on our well-being, will require understanding the historical role of information in Western society. We will need to identify the continuity of that role into which discontinuities in the form of technological innovations were introduced, changing the form of expression of that role.

For example, long before the Internet erupted into public consciousness in the 1990s, it was obvious that fundamental structural changes already had taken place in advanced industrial economies and that these changes were the result of the growing value of information. By the late 1920s, the rate of growth of the service sector of industrial economies exceeded the rate of growth of the industrial sector. Shortly after the Second World War, and for obvious reasons first in the US, the size and the value of the service sector exceeded those of the industrial sector in absolute terms, and the gap between the two widened steadily from then on. By the mid-1950s, "information workers" were the largest component of the US workforce and, by the end of the twentieth century, manufacturing accounted for less than 15 percent of the US GDP.

In 1962, the Austrian-American economist Fritz Machlup published *The Production and Distribution of Knowledge in the U.S.* (1962). Machlup argued that an emerging "knowledge economy" was replacing the manufacturing-based industrial economy. In 1973, Daniel Bell's *The Coming of Post-industrial Society* (1973) correctly predicted the following: the imminent domination of economies and of societies by information-based services; that commercial competition would be driven by information-based "intellectual" technologies; and the replacement of the machine as the central Western metaphor for nature, human beings, and organizations by information. In 1982, Marc Porat produced and narrated a highly acclaimed video documentary, *The Information Society* (1982); he then went on to a notable career as an information economy entrepreneur and in-house visionary at several influential computer hardware and software companies. In the 1984 *New York Times* best seller *Turing's Man*, University of North Carolina literature professor J. David Bolter identified the computer as the "defining technology" of the late twentieth century, displacing such earlier defining technologies as the clock, steam, and electricity.

Information in its Western historical context

By 1980, then, well before adoption of the information technologies that are ubiquitous today, an information "revolution" had been proclaimed as successor to the Industrial Revolution. An information-based "knowledge economy" was hailed as imminent successor to the century-long dominance of the manufacturing-based economy. In the process, the influence of the modern industrial corporation with its extensive social, political, and financial infrastructure was weakening. An economy dominated by information and knowledge – the close relationship between these will be left intuitive here – entailed new institutions and a new social infrastructure, and this certainly *seems* to justify talk of revolutionary social change triggered by the transition.

It is arguably the case, however, that the transition from an industrial to an information economy was evolutionary, that it was continuous with changes in Western societies that had begun with the invention of the university in the twelfth century. Furthermore, it was continuous as well, in an ironic Hegelian way, with developments internal to the Industrial Revolution itself. That is, it is arguable that the role of information in Western societies since the twelfth century, as measured by the value placed on information, has been continuous, in spite of the introduction of successive "revolutionary" information technologies. And it is also arguable that it was in the service of the maturing mass production-centered Industrial Revolution that the seed was planted for the industrial dominance of society to give way to the dominance of information and knowledge. Furthermore, that both of these continuities are correlated reinforces the *evolutionary* character of the information *revolution*.

The university as information institution The medieval university was a social invention, designed to collect information, interpret it, and disseminate it. There had been schools in western Europe prior to the mid-twelfth century, of course, including Cathedral schools for advanced study, law schools, and medical schools. There were also advanced study institutions across the Islamic world, with which Christian Europe was actively engaged. Nevertheless, the university, appearing first in Bologna and Paris in the second half of the twelfth century, was a distinctive social-intellectual institution and, as such, an invention. It was not an inevitable outgrowth of eleventh-century western European social, educational, or intellectual needs; and when it was introduced it was perceived by leaders of the Catholic Church to have a threatening heterodox character. Nevertheless, in the wake of the founding of the Universities of Paris and Bologna, a university movement arose that within 150 years saw the establishment of scores of universities from England to Poland and from Sweden to Spain.

The university was and remains an information- and knowledge-intensive institution where learning was almost exclusively book-centered. The intellectual motivation underlying the creation of the university finds expression in the secular texts that made up the curricula in philosophy – comprising astronomy, music, geometry, and arithmetic as well as logic, grammar, and dialectic – in civil law, and in medicine. These were, in the first instance, ancient Greek and Roman texts, but included more recent Arabic-language works as well. The rapidly growing university market for these texts prompted the creation of a commercial translation industry that funneled works by pagan, Jewish, and Islamic authors into the hands of Christian scholars, who studied them, interpreted them, and taught them.

The intellectual motivation behind the invention of the university also finds expression in the prosecution of Peter Abelard by Bernard of Clairvaux. Abelard was the first Western public intellectual and Bernard was a politically powerful Churchman, an organizational and managerial genius as master of the far-flung network of Cistercian monasteries, and an original theological thinker. The record of his prosecution of Abelard shows that Bernard very clearly understood that the motivation behind what would erupt as the university movement entailed the claims that reason was at least coequal to faith, that intellectual inquiry was autonomous vis-à-vis religious teaching, and that the autonomy of reason implied the value of secular knowledge *for its own sake*. Though this posed a clear threat to the mission of the Catholic Church, Bernard did not attempt to quash the newborn university. That he settled for cautioning against the teaching of heresy suggests the rootedness of the university in twelfth-century western European society (Gilson, 1955; Poole, 1960).

The book as information technology In fact, the universities that were founded shortly after Abelard's death *would* claim autonomy for reason, and the value of *all* knowledge for its own sake, independent of its value for the saving of souls. The claim to the autonomy of intellectual inquiry, and its successful defense over centuries of opposition, can be said to have been the necessary condition for the emergence of an information cum knowledge economy in the West. At any given time *before* the introduction into European society by Johann Gutenberg of printing using movable metal type *c.* 1450, tens of thousands of young men were enrolled in universities across Europe. This created a constantly expanding market for books, a market that evolved as the recovery of ancient texts stimulated the production of new editions, new commentaries on and interpretations of them, and of original works as well. It also established the book in Renaissance society as the medium of choice, not only for transmitting knowledge, but also for transmitting know-how, that is, "art," what we today mean by technology.

The Platonic–Aristotelian conception of knowledge as universal and abstract can only be embodied in a written text, if it is to be transmitted beyond a small circle of orally instructed disciples. For millennia before the invention of writing, however, and for millennia after, very sophisticated *know-how* accumulated without texts. Know-how, and the information it integrates, is embodied in artifacts and in the processes of their production. It is acquired by doing and the making is itself evidence of degree of mastery. The medieval craft guild system, surviving into the early modern era, was illustrative of this fact, yet by the fifteenth century it had become commonplace to teach know-how using books. This is symptomatic of the penetration into society of the text-centered university culture. These texts aimed at teaching such new forms of know-how as central vanishing point perspective drawing, arithmetic using Hindu–Arabic numerals, double entry bookkeeping, military engineering, deep mining, metallurgy, and the design and construction of machines and water systems.

It was into this highly knowledge-charged and book-oriented social environment that Gutenberg introduced a new printing technology that triggered Europe's first information "revolution." The details of the technology are far less important than the response to its introduction. By 1500, some two million books and some eight million pamphlets and tracts had already been printed, and the volume of printed material and the audiences for it have continued to grow to the present day. To enable this, a vast infrastructure had to be created, from collecting massive volumes of rags to meet the demand for paper, to whole new industries for creating, printing, publishing, and distributing written information (Eisenstein, 1983).

Why? Why did the new print technology evoke so different a social response from the Western societies into which it was introduced than from the Chinese, Korean, Japanese, and Islamic societies into which it had been introduced centuries earlier? The historian Lynn White (1962) argued that newly introduced technologies only open doors for a society; they do not compel entry. It follows that the response of a society to an innovation contains important insights into that society's values. The response is not dictated by the innovation; it is evoked by it. The response to Gutenberg's print innovations, for example, as later to the penny post, telephone, and Internet, reveals a deep, antecedent social value placed on sharing information. From one perspective, the value placed on sharing information is instrumental, but from another perspective, the value lies in what G. H. Mead (1967) called *communitas*: communicating in order to create communion.

The print technology-based information "revolution" occurred in a social context *already* committed to the value of information and knowledge as reflected by the status of the university and the textualization of know-how. The new technology and its supporting infrastructure enabled extension of the expression of the value placed on information and knowledge in Renaissance and early modern society. And as technological innovations invariably do, unanticipated applications emerged and unanticipated social consequences followed, which strengthens the evolution analogy. The extraordinary increase in the dissemination of information led to the creation of institutions to control piracy, heresy, sedition, libel, and falsehoods. The growing value of information and of knowledge led to the introduction of laws and agencies to protect intellectual property rights (Johns, 1998). By the seventeenth century information and knowledge were commercial "products" reflecting a flourishing industry that grew, driven by innovative *packaging* of information, for example, Lloyd's Register of Ships (Bernstein, 1996).

Extending the scale and scope of information The rise of modern science in the course of the seventeenth-century "Scientific Revolution" was from the beginning tied to texts and their dissemination. So was the "revolution" in European mathematics initiated by the adoption of Hindu–Arabic numerals and then of algebra, which over the next century displaced geometry as the "language" of European mathematics (Grattan-Guinness, 1997). In addition to major advances in the power of algebra and the invention by Leibniz and Newton of the calculus, probability theory was created. Initially concerned with applications to gambling and games, by the eighteenth century, first at the hands of Jacob Bernoulli, later by Turgot and Condorcet, probability theory was applied to questions of social, political, and economic policymaking (Hacking, 1990). This, in turn, laid the foundation for the social statistics movement in the nineteenth century that made the collection of information about individuals, businesses, and nature an increasingly valuable enterprise to governments and to businesses. By the end of the nineteenth century, it was an independent commercial enterprise as well, using new technologies to create information products and new businesses to manufacture information machinery (Porter, 1986).

In the course of the nineteenth century, growing populations and high literacy rates, together with pressure for universal, publicly funded education, were accompanied by an explosion in the quantity of information aimed at the general public in the form of newspapers and magazines. This information explosion was enabled by numerous technological innovations, among them in the production of paper from cellulose, rather than from rags, and in printing, ranging from typesetting machines to high-speed presses printing ten thousand newspapers per hour. As the US population soared from three million in 1790 to over ninety million by the 1890s, the quantity of print material increased in step and in variety. In 1783, 43 newspapers were published regularly in the new republic. By 1814 there were 346. The census of 1850 counted 2,526 newspapers, daily and weekly, and the census of 1880 counted 11,314, including illustrated weekly magazines (Starr, 2004). The introduction of national and international postal service standards and treaties, at much lower rates than before, unleashed an astonishing increase in the flow of mail. The "penny post" was introduced in England by Rowland Hill in 1839, exploiting the new railroads being built. In that year, 39 million letters were posted in the United Kingdom. The next year 168 million were posted, and in 1850 347 million. Again, the response to information-related innovations reveals a continuity in the underlying value placed on acquiring and sharing information.

The enormous growth in the volume of mail was enabled by the introduction of steam-powered transportation technologies that vastly increased the speed with which mail could be exchanged at virtually no extra cost to the carrier. On the face of it, transportation and information technologies are independent of one another, but technologies that enhance the dissemination of information by that very fact intensify its production, if the social value of information exchange is already in place. The increase in production was *enabled* by the innovation, but not caused by it, which further reinforces the evolution analogy.

New nineteenth-century communication technologies also enhanced the dissemination of information. The introduction in the 1840s of commercial electric telegraphy had an effect on the production and consumption of information comparable to the linkage of postal service to transportation innovations. Within forty years, undersea telegraph cables linked the Americas, Europe, Asia, and Australia (Standage, 1999). The commercialization of the telephone, beginning in the 1880s, added another information exchange modality. It spread rapidly, in spite of the prediction by many that not enough people had anything important enough to say to one another for the telephone to become a household necessity! In both cases, new kinds of information companies sprang up, and not just telegraph and telephone companies. An important instance of new companies that sold information as a product were those based on the stock market "ticker tape" machine, giving instant information about marketplace transactions. It was the sale of his invention of an improved ticker tape machine that gave Thomas Edison the funds he used to build his Menlo Park "invention factory," as he called it, which produced, among other inventions, the incandescent lightbulb, central electricity-generating systems, the phonograph, and moving picture cameras. With the commercialization of radio broadcasting starting in 1919, and then of television primarily from the late 1940s, every citizen of the industrial societies was routinely inundated by an unprecedented quantity of information, long before digital technologies came into play.

The information side of the Industrial Revolution The "Industrial Revolution" began with the invention of new spinning and weaving machines that introduced mass production to the textile industry. By 1800, these machines, initially hand- or water-powered, converged with James Watt's improved steam engine, which soon spread to other industries stimulating broader applications to production and to transportation. The rapid adoption of these technologies then transformed society, globally. Between 1800 and 1900 there erupted an unanticipated, unprecedented, and apparently self-sustaining flood tide of innovations in production, transportation, and communication technologies; in the commercial integration and exploitation of these innovations; and in the organization of the innovation process itself (Landes, 1969). By the early twentieth century, it was accepted as a truism that industry was the *real* source of wealth (and of power) in "modern" societies. Making *things* – bridges, dams, tunnels and roads, skyscrapers, dynamos, electrical networks, and mass production factories, automobiles, airplanes, refrigerators, washing machines, telephones, and radios – was the most valuable activity for a society and for its individual members.

At the very same time, however, recognition began to spread that the social and economic dominance of manufacturing and of industrial corporations, while obviously grounded on the commercial exploitation of physical inventions and on having physical products to sell, was critically dependent on information and on knowledge (Chandler and Cortada, 2000). This recognition was the "seed" out of which would grow the post-Second World War, pre-Internet perception that information and knowledge were supplanting manufacturing as the dominant forms of wealth-generating activity in modern economies. The information economy thus evolved out of the very nature of the manufacturing economy.

Maintaining industry's dominance was recognized in the last third of the nineteenth century as crucially dependent on the collection and assimilation of ever more information, and on the continual production of *new* knowledge. This was expressed, for example, in the industrial version of the craze for the collection of statistics that swept over "modern" governments from the early nineteenth century on. The history of accounting reveals that, during the nineteenth century, the systematic collection, recording, and analysis of detailed numerical information about the operation of businesses became a central concern of entrepreneur owners (Johnson and Kaplan, 1987). This concern accelerated with the explosive growth of industrial corporations with far-flung operations that demanded new accounting metrics adapted to these distributed operations. In Germany, for example, this was stimulated by the extraordinary growth from the 1860s on of chemical companies based on the manufacture of synthetic dyes and their many by-products, among them explosives, fertilizers, and pharmaceuticals (Travis, 1993). These companies quickly recognized their dependence on new results from chemical research in order to maintain competitive advantage by developing new products to sell and more efficient, faster, and cheaper processes for producing them (Travis and Reinhardt, 2000). The result was the creation of chemical research laboratories within chemical companies and the forging of ties with university-based chemists and their students.

In America, the spread of centrally administered, hierarchically organized, vertically integrated manufacturing corporations, pioneered by the Swift Meat Packing Company in the 1880s, made information a critical resource for managing operations and for assessing the success of management, at its multiple levels. One symptom of the importance of information to industry was the emergence and growth of the mechanical, later electromechanical, arithmetic calculator industry. A commercial calculator industry first emerged at the beginning of the nineteenth century, using versions of a calculator invented by Leibniz one hundred years earlier. By the end of the century, however, a host of companies were mass producing a wide range of new designs, reflecting the influence of the typewriter in their modes of inputting data and outputting information. "Adding machines" were as ubiquitous in business by 1900 as computers are today and the effective management of businesses large and small was dependent on the flow of the numbers produced by these machines to owners and executives (Cortada, 1993).

Manufacturing transforms the university By the last third of the nineteenth century the "high-tech" industries of the day were experiencing explosive growth: the chemical industry, the electricity generation and distribution industry, the electrical appliance industry, and the telephone, railroad, steamship, and automotive industries. It was a fact of life that the continuing profitability of these industries was crucially dependent on a continuing stream of new inventions, on the continuing evolution of the engineering knowledge necessary to convert those inventions into commercially valuable products, and on the availability of vastly greater numbers of engineers to do this work. Because of the increasingly specialized nature of the new industries, the new engineering knowledge needed was increasingly dependent on training in mathematics, physical science, and laboratory methods. It is not coincidental, then, that the growth of engineering curricula at established colleges and universities, and the creation of new engineering-specific institutes of technology, tracks the growth of industrial corporations.

The US Congress in 1862 passed the Morrill Land Grant Act rewarding state governments that founded agricultural and "mechanical arts" colleges. At that time, only six institutions offered degrees in engineering and two of those were military academies. Thirty years later, the number was 120 and it doubled to 240 by 1920. Collaterally, there were in 1860 perhaps three thousand people in the US with the job title "engineer" and only a small fraction of these were college trained. By 1960 the number was a million and it doubled to two million in 1980, by which time almost all engineers had degrees from officially accredited institutions (Goldman, 1991). This enormous expansion in the number of engineering colleges created a demand for unprecedented numbers of academic mathematicians, physicists, and chemists. Although initially employed largely to serve the needs of engineering students, increasingly professionalized communities of scientists and mathematicians quickly arose. At the same time, growing numbers of mathematicians and physical scientists were required by industry. They were needed to solve problems in factories and to develop commercially valuable new knowledge in in-house research laboratories such as those established at Dupont, General Electric, and AT&T.

Together, these professionalized communities of academic and industrial scientists, mathematicians, and college-educated engineers were high-visibility, specialized knowledge workers. The knowledge worker model expanded with the introduction of "management science" in the opening decades of the twentieth century. This brought the business world broadly into the knowledge worker domain. To be a successful corporate manager now required education and training in "rational" decision making and in financial management models. By 1919, Thorstein Veblen was calling attention to what he saw as a threatening transition in American industry. The entrepreneur-innovators running businesses they had created, and knowledgeable about the specific technologies involved in those businesses, were giving way to lawyers, financiers, and accountants with generic "scientific" management skills and little understanding of technology (Veblen, 1921).

There was, then, a "pull" from industry for creating educational institutions and degree programs capable of providing the army of knowledge workers and of information-collection and -processing workers that industry needed. The ability to recruit such an army, needed first by industry and starting in the 1930s by rapidly expanding federal and state governments, was made possible by an extraordinary social development, one in which the US again played a lead role: the extension of college education to the public at large. From the twelfth through the nineteenth centuries, a college education in the West provided entry to, or at least association with, social elites for those students from the lower or middle classes fortunate enough to acquire a college education. In the course of the first half of the twentieth century, however, a college education for everyone who completed high school became a US social ideal and then an expectation. After the Second World War, it became a social as well as a personal goal in Europe too and then globally.

The resulting expansion in the number and size of colleges, to some three thousand in the US alone, serving over three million students at any given time, required a concomitant expansion in the number of faculty, administrators, and staff: thus, more knowledge workers. More significant was the impact of this expansion on the careers that a college education opened for the large majority of new college students who were not pursuing careers in natural science, engineering, math, law, or medicine. Clearly, there was no point going to college only to wind up after graduation working as a manual laborer or as a clerk in a store. The most popular degree programs for these students were business, education, English, social science, psychology, and history. And the majority of these graduates became either teachers for the now vastly expanded kindergarten through the graduate and professional school educational system, or they became information workers. The market for such workers was in businesses across the economy, in government, in new kinds of publicly funded social-service institutions, and in rapidly expanding information businesses such as advertising, human resources services to industry, marketing, sales, and mass media entertainment. This social development was driven by the growing value to society of information, echoing the value of information to industry and its supporting infrastructure.

The Great Depression called into question the claim that knowledge guaranteed prosperity, but by the end of the Second World War knowledge was once again critical to national and personal prosperity. The media dubbed the war a "scientists' war" and lionized Vannevar Bush for overseeing the creation by teams of industrial and academic scientists and engineers of radar systems, the atomic bomb, cheap penicillin, the proximity fuse, operations research, and ENIAC (Electronic Numerical Integrator and Computer), among other wartime-developed wonders (Zachary, 1997). Information and knowledge, and the ability to exploit them productively, were proclaimed vital social and economic resources. They had suddenly become the ultimate wellsprings of individual and social security, wealth, power, and value. Bush argued just this case in his seminal report to the President of the United States entitled *Science: The endless frontier* (Bush, 1980). This report influenced the massive US federal government support of science and technology that began with the onset of the Cold War. This support, in turn, directly and indirectly influenced the creation of the semiconductor-based digital information technologies that would lead to talk of an Information Revolution.

Information trumps production Mighty manufacturing, it transpired, was symptom not cause of prosperity, power, and modernity. Manufacturing was symptomatic of the exploitation of information and knowledge, for example, via mass production for mass markets. The maturation of manufacturing revealed its dependence on information and knowledge. Innovations in collecting, interpreting, and distributing information, and in the generation and "packaging" of information and knowledge, *for industry itself* led to the creation of a growing range of information "products." By the 1960s, references to an emerging "information society" and "knowledge society" reflected a new truism: that goods and services alike were, in the end, just forms of information and knowledge. With this realization came a new metaphor, of the mind as an information-processing "machine."

What in the mid-1950s would emerge as the field of artificial intelligence research began in the 1940s with the McCullough-Pitts electrical model of neuronal networks, Alan Turing's essay on machine intelligence, and Norbert Wiener's new science of cybernetics (Heims, 1991; Wiener, 1948). The 1953 discovery of the structure of the DNA molecule led within a decade to its characterization as an information structure, so that life itself became an information phenomenon (Holmes, 2001; Morange, 1998). The focus on information was reinforced by the eruption after the Second World War of yet another flood tide of technological innovations. This time the innovations were in digital electronic information and communication technologies, the latter now explicitly recognized as channels for transferring information. The computer, hitherto conceived as a super-calculator, was reconceived as a universal information processor. Its rapid adoption by industry is what led Bolter to identify it as the "defining technology" of our age in his book *Turing's Man*. Looking back to 1984 from the perspective of today's information technologies, however, the penetration of the computer had barely begun. There was no PC and no Mac, no e-mail, "chatting," blogs, or text messaging, no Amazon, eBay, or WorldWideWeb, no Google, GPS, or MapQuest, no Facebook, MySpace, YouTube, or Second Life, no MP3 players, digital cameras, DVDs, cell phones, laptops, Palm Pilots, or BlackBerries, and no outsourcing of call centers and radiology departments to India and Australia.

Whole new industries have emerged since the 1960s devoted to creating and selling information "products" to consumers, industry, and governments. Information technologies have transformed the prevailing system of production of goods and services, as well as transforming traditional goods and services themselves into "smart" goods and services. All this was made possible by the extraordinary development of progressively smaller, cheaper, and more powerful semiconductor-based electronics technologies and of innovative industrial and consumer applications of those technologies. By the mid-1980s, computer-controlled machine tools anchored a cluster of digital information technologies that enabled large-volume production of highly individualized niche market products, as opposed to mass production commodities. The transformation of all-purpose sneakers into special-purpose sports shoes illustrates the point, as does the evolution of the telephone from a solemn, black, heavy table top appliance to colorful stylized wall and desk phones to portable cordless phones to high-fashion cell phones too small to use comfortably yet incorporating digital cameras and able to download text, audio, and video files via the Internet anywhere in the world.

The dot.com "bust" notwithstanding, the Internet continues to be an astonishing social, economic, technological, and cultural phenomenon. It is, in addition, an abstract information structure that emerged as an unintended consequence of the invention of information exchange protocols (TCP/IP) in 1974 by Vinton Cerf and Bob Kahn (Abbate, 1999). As with all major innovations, the Internet is woven into a web of associated, mutually reinforcing technologies and socially sanctioned applications with which it has coevolved and whose social impact has been amplified by this coevolution.

The impetus for networking computers came from officials at the US Department of Defense Advanced Research Projects Agency (ARPA) who wanted to monitor more effectively computer projects they funded at different universities. The success in 1969 of linking first two, then three, computers at different locations quickly led to the formation of networks of computers able to exchange information digitally. The growth rate in the number of networks, and in the number of computers in these networks, was breathtaking. Already in the 1970s the networks themselves began to be interconnected into a hierarchy of networks. This hierarchy had been dubbed the Internet by 1983, though it was better known in the US as the Information Superhighway into the mid-1990s. The formation of the Internet was made possible by the strict observance by all participating networks, and the individual computers they linked together, of the TCP/IP information exchange protocols that were developed for ARPAnet. The Internet is, therefore, truly an abstract information structure, not a thing at all in the traditional, physical artifact, Industrial Revolution sense of "thing." The impact on society at large of the Internet was utterly transformed by its convergence with the WorldWideWeb and with powerful software for finding and correlating information on Internet-linked computers.

The web was created *c.* 1989 by Tim Berners-Lee to facilitate collaboration among the hundreds of scientists, engineers, and mathematicians who make up high energy physics research teams. The academically developed Gopher file locater program evolved in 1993 into the commercial web page "browser" Netscape Navigator and into the commercial web page search engines WebCrawler (1994), Yahoo (1995), and Google (1998). E-mail had been standardized by Ray Tomlinson in 1971, and Internet-based discussion forums by Tom Truscott and Jim Ellis in 1979, but e-mail for the masses and blogging were not the objective of these highly innovative software developments, nor were eBay, retail electronic commerce, global interenterprise collaboration among companies, outsourcing and the virtual enterprise, and global pornography and gambling rings. All of these nevertheless quickly emerged as what the computer community calls "killer apps." The course that innovations take as they are applied, and hence the social impact that innovations have, cannot be deduced from attributes of the innovation!

By the turn of the twenty-first century, digital information applications of semiconductor technologies were driving the emergence of new, global social, cultural, and commercial information infrastructures with profound consequences for prevailing institutionalizations of social, political, and commercial activities, as well as for prevailing personal, social, and cultural values. While this may *seem* revolutionary, and is popularly taken to be so, even a brief historical sketch such as this one strongly suggests that the "Information Revolution" in fact is the most recent stage of a very long evolution of the role played by information in society. History here illuminates the continuities underlying the technological discontinuities shaping the expression of this role. At the same time, this sketch raises further questions about information itself. How is it that information and knowledge have emerged as the dominant drivers of social and cultural change today? How does technological innovation effect social change? To what extent has the transition from industry-dominated to information-dominated societies improved human well-being? Before the issues raised by these questions can be addressed, however, it is necessary to clarify what we mean by "information."

What does "information" mean?

There are, in fact, three different, though related, usages of "information" extant today. The first and most familiar usage is content. The second usage is for the content-independent theory underlying digital information technologies; the third, for a type of relational structure that is increasingly identified as an elementary feature of physical reality, on a par with matter and energy. In all three of these senses, information is routinely depicted as revolutionary today, but only the second seems truly revolutionary.

Information as content For most of us, to speak of the growth of information in (Western) historical context is to speak of the growth of information as content: the content of books, newspapers, journals, magazines, radio and television broadcasts, music, films and videos, web pages, e-mails and blogs, 'phone conversations, "snailmail," catalogs, and billboards. The most common reference to an information "revolution" is to the volume of information-as-content to which everyone is exposed. The already heady nineteenth- and twentieth-century rate of growth of information accelerated yet again in the 1990s and continues to accelerate as digital technologies and new applications they enable for creating and sharing text, audio, and video content via the Internet continue to be introduced.

The most common concern voiced about the rate of growth of information-as-content, a concern voiced long before the digital technology-based acceleration in quantity of content, was that we could not possibly process even the fraction of information in which we are awash that was plausibly relevant to our own interests. Content available via the Internet has intensified this concern along with the equally historical problem of assessing the accuracy of the information available to us. To these must be added the time it takes to process so much information delivered via so many modalities. As of late 2007, YouTube subscribers had uploaded over 60 million videos to its servers; MySpace had 200 million subscribers, and Facebook another 60 million; Second Life had millions of paying "residents," 65,000 of whom were actively "living" there at any given time and who created a real world-linked economy through their virtual activities. Millions of people worldwide play massively multiplayer video games, and blogs have become a potent, global social and political force. The scope and the scale of these social networking and virtual reality software applications are constantly increasing and at a rapid rate.

What makes it arguable that the social role of information in the sense of content has been continuous in Western societies since the fifteenth century is the constancy of the response to every newly opened technological innovation "door," to use Lynn White's metaphor. Successive information technology creation and dissemination innovations have resulted in new outlets for the role that information has always had. "Information" in the world of digital information technologies, however, is a meaningless statistical pattern and information in this sense has a better claim on being revolutionary.

Information as mathematical pattern At the conceptually opposite pole from information-as-content is the counterintuitive definition of information as form, independent of its meaning. It is, however, precisely this formal definition of information that underlies all of the digital information technologies that in the last twenty-five years have inspired talk of ours as an Information Age. Ironically, these very technologies to which the semantic sense of information is irrelevant have enabled the accelerating expansion of information as content.

The man behind this formal definition of information, and thus the man more than anyone else deserving of being called the "father" of the Information Age, was Claude Shannon. In 1948, Shannon, then working at Bell Laboratories, where a research team led by William Shockley was in the process of inventing the transistor, published a truly epochal paper, "A mathematical theory of communication." Shannon's theory of communication, which contains within it a theory of information and a theory of its optimal encoding, became the foundation for the digitalization of electrical, electromagnetic, and optical signals (Shannon and Weaver, 1948). That is, it became the foundation of current computer storage technologies, music CDs, video DVDs, optical fiber telecommunication technology, cell phone technology, and more.

What is truly radical about Shannon's theory of communication is that it sets aside as irrelevant the meaning of the message being transmitted. What Shannon proposed was a statistical theory of the information-carrying capacity of any communication channel, considered to be made up of a source transmitting a signal, a receiver of that signal, plus an intermediate storage or delay stage. The prevailing wisdom was that the frequency of errors in transmitting an electrical or electromagnetic signal was a function of the rate of transmission (thus, the slower the rate the more accurate). Shannon showed that error rate was independent of the transmission rate and could be reduced to an arbitrarily low figure even at very high speeds. The key was encoding the signal in such a way that redundancy mitigated the effect of errors. So, within his information theory, itself embedded within his communication theory, Shannon incorporated a theory of coding information signals in his theory of information. (Recall that communication is the dissemination of information, so the two are intimately connected.)

For Shannon, information is defined in terms of uncertainty. Information is a statistical difference between what we know and what we do not know. (A message whose content you already know contains zero information!) The problem Shannon initially set himself was how an automated receiver, with no human intervention, could reliably distinguish an incoming signal from random noise and accurately reconstruct that signal. What the signal was about, what it *meant*, was irrelevant. Communication in Shannon's theory is a stochastic process and information as content is "merely" a statistical pattern, a semantically meaningless statistical pattern. It can be a list of stock prices, a cry for help, *Hamlet*, or a Bach fugue. Content per se vanishes and is replaced by its statistical *doppelganger*!

The source portion of Shannon's problem was defined by the nature of the signal being sent and here is where the information is characterized statistically, independent of meaning. Shannon discovered that the equation correlating information and uncertainty had the same form as the equation in thermodynamics correlating entropy and the temperature of a physical system. This suggests, startlingly, that there is some connection between information and entropy. He proceeded to develop his information-characterizing equation using mathematical tools "borrowed" from late nineteenth-century statistical mechanics and from Norbert Wiener's mid-1940s use of statistics in the new science he called cybernetics.

For the channel portion of his problem, Shannon showed that the optimal balance of speed and accuracy of signal transmission for a given channel was determined by how the signal was encoded and he also showed that the optimal encoding was binary or digital. If a continuous wave signal, for example, the electrical signal generated in a microphone by a musical instrument, were chopped up into tiny "bits" very rapidly, then a perfect reproduction of the original signal is possible, at least in principle and very nearly so in practice. The original signal is digitized by converting it into a stream of binary "bits" that can be transmitted as rapidly as prevailing technologies allow – at the speed of light in optical fibers – and then stored in any binary format whatsoever, for example, as pits burned onto the surface of a CD or DVD by a laser, or as magnetized regions on a computer disk drive, as electrical pulses for cordless and cell phones, or as light pulses in optical fibers. The bit stream can be reconverted directly into the original continuous signal driving an audio speaker to reproduce the sound of the original instrument, or it can be reconstructed from the stored binary format. Again, the software that controls this process is utterly indifferent to the semantic content of the signal being digitized.

The interpretation of information as a statistical pattern independent of its meaning, and the incorporation of this interpretation into a host of technologies that pervade life today and overwhelm us with information content and communication channels, seems revolutionary rather than evolutionary. Shannon used ideas and tools derived from George Boole's pioneering nineteenth-century work in symbolic logic, from thermodynamics (making entropy a measure of the information content of a message), from binary arithmetic, from statistical mechanics, and from cybernetics. But there is no meaningful sense in which his theory of communication and information can be said to have simply emerged out of those elements.

At the same time that he worked out his nested theories of communication, information, and information coding/decoding, Shannon worked out the generic digital Boolean electrical circuits for simulating logical reasoning. This was just in time for these designs to be picked up by the first generation of designers of digital electronic computers. Shannon's logic circuit designs became the cornerstone of the logic circuits in virtually every computer to this day. So if the convergence of digital computers and information technologies is the necessary condition of the Information Age and the agents of the social changes associated with that age, no one has a better claim to being its "father" than Claude Shannon.

Information as reality The third meaning of "information" extant today is ontological, treating information as an elementary feature of physical reality. The essence of the mechanical philosophy of nature since Descartes was that all natural phenomena were to be explained solely as the results of the complex motions of matter. In the nineteenth century, without abandoning the quest for mechanical explanation, physical scientists were led to attribute physical reality to immaterial entities such as energy, fields, waves, and even to structural relationships, for example, the spatial arrangement of atoms within a molecule. What made these entities physically real was that they had causal consequences. Molecules with the same atomic composition, for example, could have different chemical and physical properties if their atoms were arranged differently in three-dimensional space. Energy, fields, waves, and relational structure are all said to be real because they have causal consequences. They are said to be elementary because they cannot be reduced to material entities. The claim that information, too, is a source of lawful causal action puts information on a par with energy, fields, waves, and structure as physically real and elementary. Indeed, early in the twenty-first century, it was being claimed that information was the ultimate "stuff" of the universe (Baeyer, 2003).

Support for the reality of information as an elementary feature of reality came from the concept of a genetic "code" following the discovery of the double helical structure of the DNA molecule. By the 1930s, DNA and RNA had been isolated as the molecules in the nucleus that fifty years earlier had been suggested as playing a central role in reproduction and heredity, probably because of their complex structure. It was simultaneously discovered, however, that DNA always contained the same four bases. Whichever life forms DNA samples were drawn from, these same four bases appeared. This led many to conclude that DNA could not be all that important a factor in determining heredity, which distinguishes life forms from one another, if it were the same in all life forms. This conclusion reinforced the prevailing view that proteins were the key to genetics, but Oswald Avery's 1943 experiments, repeated in the early 1950s, strongly suggested that somehow DNA was key (Morange, 1998).

In 1953, Watson and Crick published their claim that the structure of the DNA molecule was a double helix whose strands were coupled by complementary pairs of the four bases that had been discovered in the 1930s (Watson, 1968). In 1957, Meselson and Stahl showed experimentally that DNA molecules were reproduced by the strands "unzipping" from one another while forming complementary strands and thus forming two new DNA molecules (Holmes, 2001). The question still remained of how the DNA molecule functioned, how it did all the work in the cell that it needed to do. In a move reminiscent of Shannon's in his theory of information, the physicist George Gamow wrote to Francis Crick suggesting that the answer to this question was coding. He proposed that it was *the sequence* of the four bases within the DNA molecule that was fundamental. The sequence constituted a code for the manufacture of proteins out of amino acids. By making certain assumptions, Gamow argued that the four bases taken three at a time would code for just twenty amino acids, and it was known that all proteins in fact are built up out of just twenty of the larger number of possible amino acids (Morange, 1998).

It follows that the key to life is a code and DNA molecules are information structures. Chemists in the mid-nineteenth century discovered that the properties of molecules are not wholly determined by the atoms they comprise, because some properties are determined by the spatial arrangement of the atoms within the molecule. Analogously, the properties that allow the DNA molecule to orchestrate the cellular processes that make each life form what it is are not a material *substance* unique to each life form but an immaterial *sequence* using the same "stuff" that all life forms use. Life is, in the end, information, an embodied coded message (Loewenstein, 1999). Furthermore, the proteins whose manufacture the

DNA orchestrates play the roles in cell metabolism that they do because of their form, the distinctive way in which each is folded. Again, structure, which like information is a form and thus immaterial, has causal properties of its own independent of its constituents, which are a means to the material realization of the form. Indeed, if different materials could be organized into a molecule with the same form as a particular protein, then that molecule would function the same way in the cell that that protein did.

It is not only in biology that information structures have come to be identified as elementary features of reality. In contemporary astrophysics, physicists use information theory mathematics to describe black holes. Like Shannon, the physicists, too, connect the thermodynamic concept of entropy with information, and in 2005 Stephen Hawking conceded that black holes are most accurately described as information-preserving structures, after a decade of vigorous opposition to that claim. They preserve all the information about the matter and energy they suck in even as that matter and energy "disappear" into the black hole.

This information, and the distinguishing characteristics of a given black hole, are a function of its surface area, not its volume as had been expected. It can, in principle, be retrieved from a full account of that surface. This has given rise to a speculation, called the Holographic Principle, that the universe as a whole is an information structure, analogous to the surface area of a black hole and to the holograms with which we are familiar: two-dimensional artifacts containing recoverable three-dimensional content (Smolin, 2001). The further idea that all physical objects, at all levels of physical reality, are information structures is being pursued by scientists and mathematicians in various disciplines.

Stephen Wolfram's *A New Kind of Science* (2002) argues that the universe and everything in it are information structures analogous to the software "objects" that emerge unpredictably within computer programs that repeatedly iterate a simple set of rules. Wolfram claims that this should be the basis for a new physics in which natural objects and their behaviors are explained in terms of a fusion of information theory, computational science, and chaos/complexity theory. Wolfram's book is highly idiosyncratic and controversial, but there are prominent academic physicists, among them Max Tegmark at MIT and Anton Zeilinger at the University of Vienna, who champion the view that the universe and its contents are information structures. Among mathematicians, an extension of Shannon's work has evolved into a respected research specialty called algorithmic information theory. This integrates information theory and probability theory with complexity theory and the theory of self-organizing systems to yield information structure representations of material objects and their properties (Baeyer, 2003).

The idea that information has an ontological as well as a semantic character, that information is or may be the ultimate "stuff" of physical reality, is thus alive and well in current scientific theory, however odd it may seem to nonscientists to put information on a par with matter and energy as "stuff" out of which a universe can be made!

One revolution, two evolutions The question of what "information" means turns out to have a complex answer. As content, there is no doubt that we are overwhelmed today by more information in more forms than any generation of human beings has had to contend with; and the amount keeps increasing. This *feels* like a revolution, but it seems more accurately the result of an ongoing evolutionary process. As such, an understanding of the role of information in society today requires a historical perspective. This is not the case with what "information" means in information theory. It is as the basis of the digital information and communication technologies built on Shannon's stochastic model of information that a true information *revolution* can be said to have taken place in the twentieth century. Understanding the social role of information in its Western historical context does not enhance our understanding of Shannon's theory (though it does bear on the rapid application of it).

That "information" refers to a fundamental constituent of physical reality is conceptually the most radical of the three usages of "information," yet it is the easiest to defend as evolutionary! This ontological usage is a direct outgrowth of the growing recognition across the physical, life, and social sciences since the mid-nineteenth century that relationships are physically real, though immaterial, causal agents. In chemistry, the spatial relationships of atoms within molecules have consequences for the way molecules interact with other molecules as well as with light, electricity, and magnetism. In twentieth-century theories in the sciences of sociology, cultural anthropology, linguistics, and economics, relationships have become central causal parameters of theories. In the special and general theories of relativity, space and time are relationships, not entities in their own right as they had been since Newton. As Shannon showed, information is yet another type of relational structure with properties that have causal consequences. For the ontological meaning of "information," then, all material objects, up to and including the universe as a whole, are a particular kind of relational structure: the one called "information." The ontological definition of "information," however revolutionary it may seem, thus finds its place in a century-long, evolutionary process of attributing reality to relationships.

Innovation and human well-being

The three distinctive meanings of "information" have in common one phenomenon with respect to the social impact of information: technological innovation. How are technological innovations able to cause social change? Are the changes driven by these innovations improving human well-being? These two questions are closely related and to answer the latter it is necessary first to address the former.

Innovation and social change Contrary to many depictions of technology by critics of its consequences, technological innovation is endogenous to society and to the human condition, not exogenous to them. On the exogenous view, technology is not itself a social process but a force that acts on societies from outside those processes, as if it had dropped out of the sky. This view is at the heart of Jacques Ellul's very influential book *The Technological Society* (1970). Ellul was responding to what he argued was an insidious threat to humanness posed by the embrace of values essential to implementing modern technologies, above all the values attributed to rationality and efficiency.

The view that technology is exogenous to society equates innovation with invention conceived as a creative act. So understood, inventions erupt, unpredictably, out of nowhere so to speak, and present themselves to society as a fait accompli. While almost all inventions disappear without a trace, and some survivors are assimilated into the social and commercial status quo, sometimes someone invents a device, or a process, that so insinuates itself into the life of a society that, feeding opportunistically off its institutions, resources, and vulnerabilities, it transforms the host society from within (Smith and Marx, 1994). Familiar instances of such inventions cited in support of this view include Gutenberg's printing press, Abraham Darby's process for using coke rather than charcoal to make iron, James Watt's steam engines, Michael Faraday's dynamo, Samuel Morse's telegraph, William Perkin's synthetic dyes, and Thomas Edison's lightbulb and central electricity-generating stations (Mumford, 1934, 1967).

Innovation versus invention Invention, however, is not at all the same as innovation. What the historical record makes clear is that innovation is a process in which inventions are selectively exploited, and this selective exploitation is a social process. Which inventions are selected for introduction into society and which are ignored, what form the selected inventions take, and how societies respond to them are all social phenomena in which the inventor's intentions are of little or no consequence. Further, it is society's response to innovations that determines the resources applied to the form of the further development of an innovation after its introduction and to the rate of its development. Both form and rate of development are unpredictable *because* they emerge out of interaction with their social context (Goldman, 1991).

Technological innovation is thus best understood as dependent upon, but distinct from, invention. Innovation *is* a social process, hence endogenous to society. It is a process in which the technical knowledge of engineers and scientists is selectively applied to advance the agendas of commercial and governmental institutions that believe they can turn the exploitation of inventions to their advantage. An invention may be the work of an individual, and the technical knowledge necessary to bring the invention to the marketplace may be in some sense objective. The criteria for selecting what knowledge will be applied, in which ways, to the further development of an invention are *not* properties of the invention. They reflect an interaction between society and its institutions. It is not the case that "things are in the saddle and ride mankind," as Emerson put it in his "Ode, inscribed to William H. Channing." Instead some people are in the saddle and are using things to enable them to ride over mankind. Furthermore, who gets to ride and which things they choose to ride and are allowed by their society to ride are a function of prevailing social institutions and values (Goldman, 1990).

Artifacts and artifactual processes silently, and largely invisibly, embody a web of social choices. Printing with movable metal type had a totally different impact on the Christian societies of western Europe in the fifteenth century than on the Chinese, Korean, Japanese, and Islamic societies in which it was introduced centuries earlier. The same is true for the impact of gunpowder and of the application of mathematics to practical problems in mapmaking, navigation, military engineering, and machine design. Japan provides a telling instance of this differential social response to what was ostensibly the same innovation. The Tokugawa Shogunate closely restricted the use of firearms so as not to undo the status of the Samurai class. The first daily newspaper began publication in 1870, two years after the Meiji Restoration and more than five hundred years after the introduction there of printing using movable metal type. The rise of a commercial nuclear power industry in the US and UK in the 1960s is incomprehensible without understanding the political and military agendas that pushed utilities to order the first generation of power reactors. The impact of the IBM PC, certainly an epochal artifact, was not the result of its superior technical performance but of IBM's place in corporate culture and the decision by IBM management not to copyright the software controlling the lowest level of machine operations. This decision, later much regretted, opened the door to the PC clones and created the mass market for computers and their applications (Goldman, 1991).

Changes versus improvements In short, technological innovation is an intensely value-laden process in which people, acting individually and through social relationships and institutions, play a profoundly influential role. This role is the source of the unpredictability in principle of the social impact of an invention or of the forms that an innovation takes over time. Any attempt to assess the impact of innovation on human well-being must take this into account. Although it is widely – and invariably favorably – quoted, the last of Karl Marx's *Theses on Feuerbach* is simply wrong and deeply misleading. Marx wrote in the last of these theses: "Philosophers have hitherto only *interpreted* the world in various ways; the point is to *change* it." The point, surely, is *not* just to change the world, but to *improve* it. The history of the past two hundred years especially is a record of repeated and continuing *changes* to the personal, social, political, cultural, and physical worlds wrought by serial innovations in production, communication, health, transportation, information, agriculture, construction, and entertainment. It is by no means clear, however, that these changes have *improved* the world, that they have made people, society, or nature *better*, though it is clear enough that they have changed them!

There is an essential asymmetry between "change" and "improve." Every improvement necessarily implies a change, but not every change implies an improvement! To make a thing or a situation better, you have to change it, but it is trivially obvious that the reverse is not the case. Furthermore, it seems that very different criteria are required for determining that a change is an improvement depending on whether we are talking about artifacts or about nature. On the face of it, there are uncontroversial – call them objective – grounds for judging that a change in an artifact or a production process has improved it, because the artificial is intrinsically intentional, that is, it is a means to a specific, explicit, end. The intentional is thus also intrinsically valuational. Everything we make or do explicitly or implicitly attributes values to ends, to means, and to the relationship between them.

The ends of human action, the purposes for the sake of which we do what we do or make what we make, always entail contingent choices. They could be different. The means we use to effect those ends are almost never necessitated by the ends, but are also contingent choices. The subjectivity of the artificial notwithstanding, it seems straightforward to make objective judgements as to whether changes constitute improvements. *Given* value assignments to an end, a change in the means either does the job better or it does not and that would seem to be the end of the matter. The design parameters for a digital camera, compact-class automobile, or raincoat tell us what it is for. You may not approve of its purpose, but it seems straightforward to justify a claim that a new model camera, automobile, or raincoat is better at accomplishing it.

Without, for the moment, considering whether objective criteria really do exist for determining that changes in an artifact have improved it, it is clear that no such criteria exist with respect to natural objects and processes. Assessing claims that a change to a natural object or process has improved it requires agreement on what the purpose is of nature as a whole, as well as of constituent processes and objects. Nature, however, is, de facto if not de jure, nonintentional and, thus, not intrinsically valuational. Straightening the Mississippi River, improving the yield of dairy cows, or genetically modifying food crops can only be said to have improved them relative to human-imposed criteria, not in and of themselves. Furthermore, earthquakes, tidal waves, volcanic eruptions, drought, and hurricanes can only be called "bad" relative to their impact on humans. For that matter, the destruction of the earth itself, or of all of the life forms on it, seems eminently consistent with a universe full of exploding stars, black holes, and colliding galaxies.

Assessing well-being That said, even judging improvements in artifacts is not as clear cut as it may seem because of disagreements about means. Many variables are involved in determining means–end relationships, some highly visible, some obscure. In addition, we often subordinate assessment of means to a focus on performance. If a change in an artifact effects a clear performance improvement relative to widely accepted criteria, we are likely to judge the change to be an improvement. We are less likely to ask whether changes in the production process to achieve the performance increase had negative environmental consequences or required exploitative labor practices in Third World countries to keep the price constant. Furthermore, technological innovations may enable ends never possible before because of means never possible before. This raises the question of whether to pursue such ends just because we have the means to pursue them, or whether to employ such means given their social and environmental costs. These questions cannot be answered in terms of the intentional structure of the relevant artifact or process. They require additional value judgements every bit as subjective and controversial as judgements about natural objects. This is manifest when assessing changes in social, political, or cultural institutions or processes. While institutions and their associated processes are, like all artifacts, essentially intentional, their goals and the values attributed to means for achieving them are so ambiguous that controversy over whether a change has improved them is inescapable.

The question, then, of whether digital information technologies have improved human well-being needs to be approached from two directions. It is relatively easy, if we focus on performance, to identify a set of objective performance criteria with respect to which digital information technologies have improved. These criteria are part of the intrinsic intentionality of a given implementation of a specific technology. Having accepted the objectives of an artifact, we can say that a later implementation is better than earlier ones because it is faster, cheaper, smaller, lighter, has more features, more memory, higher resolution, greater bandwidth, more channels, brighter screen, better sound. The new implementation is judged to be better because changes can be measured against criteria that have antecedently been assigned positive or negative value (slower, heavier, more costly . . .).

Including in our assessment of changes to an artifact the means employed to accomplish them complicates the matter. We may find that we must import criteria from outside the strictly technological in order to determine that changes in the means employed to increase performance, or to lower cost, support the judgement of improvement relative to the *total* means–end system. The matter is further complicated if we need to assess whether a totally new technology-enabled capability has improved human well-being. To judge that human well-being has improved as a result of any changes to the human condition collapses into the problem of specifying what human beings, and human life, are for. Certain criteria seem relatively unexceptionable, for example longevity, morbidity, infant mortality, income. Defining physical and mental health, however, or what is normal, physically, emotionally, and intellectually, has proved intractable. Defining happiness objectively, in a universally acceptable way, seems hopeless, especially given the lack of convergence after thousands of years of effort on a single theory of ethics that would assign universally accepted values to choices.

How can we judge that human life is better, or worse, because of the computer and the web of information technologies and applications in which we are irreversibly entangled today? Was it better or worse one hundred years ago for young people to read books for hours at a time instead of socializing, playing games, or exploring nature? Was it better or worse two and three generations later for young people to listen to radio or watch television or surf the web for hours at a time instead of reading books, socializing, playing games, or exploring nature? Criteria *can* be identified for making these judgements, as for judgements that digital information technologies have improved human well-being or harmed it. These criteria, however, and the judgements that derive from them, are inescapably contingent upon subjective assignments of value to ends, and to particular means for accomplishing those ends. Such assignments of value are woven into the fabric of personal and social life.

Individually and collectively we pay a difficult to calculate but unquestionably very high "price" – measured by social, political, cultural, and environmental change, as well as economically and physically – for the mobility conferred by the automobile and for the prosperity generated by the Industrial Revolution. The price of adopting digital information technologies is not yet clear, because the impact of these technologies is still evolving.

Furthermore, what we mean by human well-being is also evolving, coordinate with the unpredictable-in-principle forms that technological innovations take as societies assimilate them. We are faced, then, with a psychological and sociological analog of the three-body problem in physics. Since Isaac Newton it has been known that there are only approximate, but no unique, general solutions to the motions of three (or more) mutually interacting bodies. In our case, technological innovation, social change, and human value judgements, including our self-conception, mutually interact in evolving contexts of creating and sharing information.

Digital information technologies, based on the revolutionary sense of information as a semantically neutral pattern, serve the evolutionary senses of information: as semantically rich communication in the service of *communitas*, via blogs, social networking software, virtual communities, multiplayer games, et cetera (Mead, 1967), and as the ordering principle of physical reality, life, the cosmos. Information technologies enable new forms of relationships, then, such that our understanding of human well-being becomes a shifting function of how these technologies affect human beings in the social and natural worlds. As they enhance the historically high value placed on information-based relationships, it is not surprising that these technologies are judged to improve human well-being. However, relationships create vulnerability to abuse and to harm, and the exploitation of the globalization of communication is causing personal and social dislocation, so while information may *want* to be free, the assimilation of information technologies is costly.

References

Abbate, J. (1999). *Inventing the Internet.* Cambridge, MA: MIT Press.

Baeyer, H. C. von (2003). *Information: The new language of science.* Cambridge, MA: Harvard University Press.

Bell, D. (1973). *The Coming of Post-industrial Society.* New York: Basic Books.

Bernstein, P. L. (1996). *Against the Gods: The remarkable story of risk.* New York: Wiley.

Bolter, David J. (1984). *Turing's Man.* Chapel Hill: University of North Carolina Press.

Bush, V. (1980). *Science: The endless frontier.* Washington, DC: National Science Foundation.

Chandler, A. D. Jr., and Cortada, J. W. (Eds.) (2000). *A Nation Transformed by Information.* Oxford: Oxford University Press.

Cortada, J. W. (1993). *Before the Computer.* Princeton: Princeton University Press.

Eisenstein, E. (1983). *The Printing Revolution in Early Modern Europe.* Cambridge: Cambridge University Press.

Ellul, J. (1970). *The Technological Society* (J. Wilkinson, Trans.). New York: Knopf.

Friedlander, W. (1957). *Mannerism and Anti-mannerism in Italian Painting.* New York: Columbia University Press.

Gilson, E. (1955). *A History of Christian Philosophy in the Middle Ages.* New York: Random House.

Goldman, S. L. (1990). Philosophy, engineering and western culture. In P. Durbin (Ed.), *Broad and Narrow Interpretations of the Philosophy of Technology* (pp. 125–152). Dordrecht: Kluwer.

(1991). The social captivity of engineering. In P. Durbin (Ed.), *Critical Perspectives on Nonacademic Science and Engineering* (pp. 121–145). Bethlehem, PA: Lehigh University Press.

Grattan-Guinness, I. (1997). *The Norton History of Mathematics.* New York: Norton.

Hacking, I. (1990). *The Taming of Chance.* Cambridge: Cambridge University Press.

Heims, S. J. (1991). *The Cybernetics Group.* Cambridge, MA: MIT Press.

Holmes, F. L. (2001). *Meselson, Stahl and the Replication of DNA.* New Haven: Yale University Press.

Johns, A. (1998). *The Nature of the Book: Print and knowledge in the making.* Chicago: University of Chicago Press.

Johnson, T., and Kaplan, R. S. (1987). *Relevance Lost.* Cambridge, MA: Harvard Business School Press.

Landes, D. (1969). *The Unbound Prometheus: Technological change and industrial development in western Europe from 1750 to the present.* Cambridge: Cambridge University Press.

Loewenstein, W. (1999). *The Touchstone of Life: Molecular information, cell communication and the foundation of life.* Oxford: Oxford University Press.

Machlup, F. (1962). *The Production and Distribution of Knowledge in the U.S.* Princeton: Princeton University Press.

Mead, G. H. (1967). *Mind, Self and Society: From the standpoint of a social behaviorist* (Charles W. Morris, Ed.). Chicago: University of Chicago Press.

Morange, M. (1998). *A History of Molecular Biology* (M. Cobb, Trans.). Cambridge, MA: Harvard University Press.
Mumford, L. (1934). *Technics and Civilization.* New York: Harcourt, Brace and Company.
(1967). *The Myth of the Machine*, vol. I. New York: Harcourt, Brace and World.
Poole, R. L. (1960). *Illustrations of the History of Medieval Thought and Learning.* New York: Dover.
Porat, M. (1982). *The Information Society* [video]. Washington, DC: Public Broadcasting System.
Porter, T. M. (1986). *The Rise of Statistical Thinking 1820–1900.* Princeton: Princeton University Press.
Shannon, C. E., and Weaver, W. (1948). *The Mathematical Theory of Communication.* Champaign, IL: University of Illinois Press.
Smith, M. R., and Marx, L. (Eds.) (1994). *Does Technology Drive History? The dilemma of technological determinism.* Cambridge, MA: MIT Press.
Smolin, L. (2001). *Three Roads to Quantum Gravity.* New York: Basic Books.
Standage, T. (1999). *The Victorian Internet.* New York: Walker and Company.
Starr, P. (2004). *The Creation of the Media: The political origins of modern communication.* New York: Perseus Books.
Travis, A. S. (1993). *The Rainbow Makers.* Bethlehem, PA: Lehigh University Press.
Travis, A. S., and Reinhardt, C. (2000). *H. Caro and the Creation of the Modern Chemical Industry.* Dordrecht: Kluwer.
Veblen, T. (1921). *The Engineers and the Price System.* New York: Viking.
Watson, J. D. (1968). *The Double Helix.* New York: Atheneum.
White, L. (1962). *Medieval Technology and Social Change.* Oxford: Oxford University Press.
Wiener, N. (1948). *Cybernetics, or Control and Communication in the Animal and the Machine.* New York: Wiley.
Wolfram, S. (2002). *A New Kind of Science.* Champaign, IL: Wolfram Media.
Zachary, G. Paschal (1997). *Endless Frontier.* New York: Free Press.

4 Work-related technological change and psychological well-being

Michael P. O'Driscoll, Caroline Biron, and Cary L. Cooper

The term "technology" has various meanings, including the windows and walls surrounding workplaces, advanced automation, computer systems, and information communication technology such as cell phones, e-mails, voice mail, personal data assistants (PDAs), pagers, BlackBerries, and the Internet. Technology is omnipresent in today's society and all aspects of our work and private lives require the use and influence of some type of technology. Over the past twenty years or so, the rate of development and implementation of new technologies in organizations and society in general has escalated, as has the rate of technological change itself. Mamaghani (2006) observed that prior to the 1990s the utilization of advanced technology was typically limited to manufacturing and production companies, with much lower rates of utilization in the nonproduction sector (such as offices). In recent times, however, there has been an explosion of technological change globally, particularly in computer-based and other media-based technologies. These are often referred to collectively as information and communication technology (ICT), whose increased sophistication and power has generated significant benefits for individuals, organizations, and society at large (Mamaghani, 2006).

Technologies have major implications for work environments and conditions, private life, and social relationships. Computerized environments and ICT provide workers with new opportunities to capture, store, analyze, and distribute information. They also increase workers' flexibility by creating mobile working practices and instant information transmission. However, this mobility, easy and rapid access to information, and modification of work processes also has profound effects on job demands, for example on the amount of time spent using technologies,[1] the amount

[1] The *New York Times* reports that "on average Britons spend 23 hours a week on the Internet, according to the Internet Advertising Bureau . . . Americans use their computers on average of 14 hours a week, according to Neilsen Media Research" (Khor and Marsh, 2006).

of information to be processed, and the training required to adapt to continuous technological change.

There is corroborative evidence that these advances will reduce the need for traditional office settings, resulting in a more mobile workforce and increased use of telecommuting (see Novaco and Gonzalez, this volume). Some argue that this increased flexibility can have positive benefits for individuals endeavoring to balance work and family responsibilities, and for organizations in maintaining a more flexible workforce which is able to respond more effectively to changing demands generated by globalization of the business environment.

There is also evidence that ICT usage is becoming pervasive in work organizations, with a greater percentage of workers relying on this kind of technology to perform their jobs (as well as for nonwork activities), compared with their counterparts twenty years ago. For example, Gustafsson *et al.* (2003) reported that approximately one-third of the Swedish working population spent 50 percent or more of their work time utilizing a computer, and around 80 percent of the Swedish population had access to cell phones and similar devices. In the US, Lazar and his colleagues (2006) presented figures from a 2001 survey indicating that 57 percent of workers used a computer in their work. Similar rates also apply to usage of cell phone technology. Indeed, in 2008 approximately half the world had access to cell phones, compared with 12 percent in 2000 (International Telecommunication Union, 2008). This illustrates how the use of ICTs inside and outside of the workplace has become pervasive in developed countries.

Despite these trends, and the obvious advantages (for both individuals and organizations) of these technologies, serious concerns have been voiced about the potentially damaging impact that ICT usage can exert upon the well-being of individual workers. These concerns have generated an upsurge of empirical research on the negative outcomes of rapid technological change on individuals, a phenomenon which has been labeled "techno-stress" (Thomee *et al.*, 2007).

In the present chapter we review the literature on the relationship between technological changes and individual worker well-being. Our primary focus is on ICTs, since these are (as noted above) the most prominent and pervasive forms of technology now utilized by workers. Also, we concentrate in this chapter on the impact of ICT in terms of *psychological* health and well-being, rather than physical manifestations. This is not to underplay the effects of technology on physical health, which have been well documented elsewhere (see, e.g., Beckers and Schmidt, 2003; Coovert and Thompson, 2003; Coovert *et al.*, 2005), but simply to enable a more focused discussion of the psychosocial aspects of technology implementation and utilization.

Our discussion centers on the effects and implications for *individuals* rather than teamwork in organizations, which has also been the subject of research in relation to new technologies. For example, in many organizations teamwork has become "virtual," utilizing an array of videoconferencing and other (typically computer-based) interactions. There is now considerable evidence that the welfare of teams in organizations is an important consideration in the implementation of new technology (see, e.g., Marshall *et al.*, 2007). Although we will not address them extensively in this chapter, these effects are also salient to the relationship between technological changes and worker well-being.

Finally, in this chapter we do not focus on the impact of technology usage at work on people's lives outside of work. This is clearly a very important issue, which is discussed at length by Kossek, Lautsch, and Eaton (this volume). Kossek *et al.* present data illustrating that the extent to which individuals feel they have control over the flexibility of their work schedule is a critical determinant of several affective and behavioral reactions, in particular the amount of conflict experienced between work and family life. We later refer to the importance of perceived control, and Kossek *et al.*'s findings confirm the centrality of this variable.

We begin by briefly describing some advantages of ICT for workers and organizations. This is followed by a depiction of how ICTs influence the workplace and, more specifically, the psychosocial environment. We then give an account of how ICT blurred the boundaries between work and personal life and report on some impacts of this phenomenon on individuals. The subsequent sections describe how ICTs influence psychological health and review studies on end-users' reactions to technologies. We also discuss how individuals and organizations can develop strategies to manage the technological challenges confronting them effectively and conclude the chapter with some recommendations for future research in this area and practical initiatives to facilitate more effective usage of ICT and enhancement of positive benefits.

Overview of the benefits of ICT

Coovert and Thompson (2003) have concisely documented changes in the implementation of technology in work organizations over the past forty or more years. They observed, for instance, that desktop computers (PCs) were first introduced by IBM in 1981, and that by the 1990s there was widespread usage of PCs and networking had advanced considerably. Networking (between technologies) has clearly had a major impact on their effective utilization, and has led to even further advances in the applications of technology to handle work tasks. In particular, technological interconnections have enabled extensive intercommunication between organizations, both "real" and virtual, which has expanded the range and volume of information available to workers and organizations, as well as accelerating the ability to perform various tasks efficiently and effectively.

Recent years have seen both evolution and revolution in ICT design and utilization. As noted earlier, there has been a significant increase in the use of various forms of ICT in workplaces, with PCs, laptops, cell phones, BlackBerries, and other computing devices now being commonly used globally. There has also been a move toward greater technological "mobility," enabling workers to fulfill task requirements in a variety of locations and time periods, rather than being restricted to a single setting (e.g., the office) and time period. This trend toward greater flexibility is likely to continue (Kostakos *et al.*, 2005), especially given organizational imperatives to adapt rapidly to changes in global conditions (e.g., currency fluctuations, market demands for products).

The benefits of such changes, for both individuals and their organizations, have been well documented (Vieitez *et al.*, 2001). There is no doubt, for instance, that technological advancements have enhanced organizational efficiency and productivity (Mamaghani, 2006), which have provided significant benefits for society as a whole. For individuals, too, more powerful and effective technology has enhanced their capacity to perform their work roles, achieve work-related goals, and (at least in principle) reduce anxiety due to lack of goal achievement (Vieitez *et al.*, 2001). In practice, however, despite the promise of better quality of work life, research has illustrated that for many (if not most) workers there has been increased anxiety, frustration, and strain resulting from their interactions with new technologies. Furthermore, it has been suggested that technology can provide opportunities for increased flexibility in terms of work hours and workplaces but, as illustrated by Kossek *et al.* (this volume), these benefits are not always realized. In the remainder of this chapter we outline the types of technology which have been focused on in research on well-being, describe some of the manifestations of technostress, including research carried out on the experiences of individual users, and discuss some of the implications of technological change for the overall health and well-being of individuals within organizations.

Influence of ICT in the workplace

According to Clegg (1994), the development and introduction of information technology in workplaces has been technology centered. Indeed, Clegg pointed out that the development and implementation of computer technology often has precedence over consideration of its psychosocial impact, which may be considered only after the system has become operational. Due to the promise of increased productivity, customer services, and integration of various information systems (Kompier, 2006), technological changes have been taken more seriously by organizations than attempts at reducing psychosocial risks associated with these changes.

Hamborg and Greif (2003) reviewed the results of cross-sectional and longitudinal studies on new technologies and stress. They noted that there is "no definite support for the simple assumption that the computer technologies *per se* might exert stress" (2003: 220). Instead, they suggested that stress symptoms are mostly found along with malfunctions of systems, poor ergonomic designs of hardware, software problems, lack of proper training prior to implementation, and inadequate implementation procedures. Smith *et al.* (1999) suggested, however, that computers not only produce indirect effects mediated by job design, but also direct effects on worker stress. They pointed out that the introduction of computers in workplaces was correlated with changes in work processes, work content, social relationships, and management style, which in turn often had deleterious effects on physical and mental health. According to a study by Carayon-Sainfort (1992), problems with computer usage are related to greater perceived work pressures and feelings of reduced control, which are in turn linked to stress. This research on office workers aimed to evaluate the effect of computer problems and computer use intensity on employees' perceptions of their tasks and their stress levels. The results showed that an increase in computer problems, such as breakdowns, and in computer use intensity, was related to an increase in perceived work pressures and to lower perceived job control.

Another aspect affected by ICT is the traditional office and the social relationships it offers. As network coverage increases and allows people to be contacted anywhere, the traditional office no longer operates as it used to. Mobile technology, video calls, and virtual meetings mean that people are present even though they are absent. As described by Introna (2001), working together no longer means sharing a time and place. Work can now be accomplished during a train journey, in a coffee shop, or at an airport (Harrison *et al.*, 2004). The type and frequency of social relationships at work are much affected by ICT. Indeed, studies in communications research suggest that physical proximity increases communication encounters (Armstrong and Cole, 1995). Indeed, when people are closer, they tend to communicate more frequently and in more situations. Though the relationship between distributed workplaces (e.g., telecommuting, virtual teams) and social isolation is complex (Novaco and Gonzalez, 2008), studies in the stress field have constantly insisted on the importance of social support for workers' mental and physical health (see, e.g., Bliese and Britt, 2001; Johnson and Hall, 1988; Johnson *et al.*, 1989; Melchior *et al.*, 2003). Mamaghani (2006) pointed out that the mobility of the workforce implies that organizations will have to adapt and find new means of protecting their health and safety.

Given the close links between ICT and the psychosocial work environment, Smith *et al.* (1999) proposed a model of stress and computer use. Their model suggests that, while some negative aspects of introducing new technologies cannot be avoided, organizations can at least attempt to minimize sources of stress in the psychosocial environment by ensuring that employees participate in the implementation process, are properly trained to use the technology, and that the new technologies do not undermine people's feelings of personal control, nor cause excessive workload. This suggestion is in line with Korunka *et al.* (1993), who studied different levels of employee participation in the implementation of new technologies and their subsequent levels of stress. Their results showed that participation in the implementation process counteracted negative effects of the new technologies.

This brief description of the influence of ICT on job design (and in turn on stress and well-being) suggests that the way ICT is implemented in workplaces and the possibility for workers to participate in the process can determine the impact ICT will have on well-being. This is consistent with stress intervention research. Indeed, a recent study by Nielsen *et al.* (2007) evaluated the impact of workers' participation in an intervention that aimed at modifying aspects of their work environment. Having influence and opportunity to shape the intervention (e.g., its content, timing, target population) mediated the relationship between information about the project and participation in the intervention activities. According to these results, employees actively evaluate their opportunities to influence the project before deciding if they will buy in and participate. This is consistent with research on job control that recommends the active participation of individuals in the intervention. For example, the widely cited study by Bond and Bunce (2001) showed that job control acted as a mechanism or mediator by which work-oriented interventions improved sickness absence and other stress-related outcomes.

The impact of ICT on workers' well-being is likely to depend on the way technology is introduced in the workplace and the extent to which it influences the psychosocial work environment. Wastell and Newman's (1996) research on the effects of a new computer system for emergency service is another example of a technological change where the new system was designed with explicit attention to human and organizational factors and strong involvement of end-users in the process. Their study evaluated the impact of a database linked with radio calls that replaced paper documentation. The new system was implemented to help ambulance dispatchers in Manchester (UK) to coordinate the command-and-control center. The results showed that, compared with preimplementation, stress and performance levels at postimplementation were improved and the system was well liked by end-users. As described by Wastell and Newman, the technical merits of a system are not enough to ensure the system will not be rejected by users. As with the introduction of changes aimed at reducing stress at work, workers need to participate and be involved in the process in order for the changes to produce intended results. In an example of good practices in stress prevention, Jordan *et al.* (2003) described how new technologies, if implemented properly and with users' involvement, can help employers provide work environments and job designs that allow higher levels of job control. In sum, new technologies are often designed and introduced in workplaces with the main focus put on their technical properties, whereas there is a growing body of research showing that, by taking into consideration humanistic principles, new technologies can actually improve well-being and job performance (Harrison *et al.*, 2004; Kendall and Kendall, 1993; Wastell and Newman, 1996).

ICT and boundaries between work and personal life

In the previous section, we argued that, if properly implemented in workplaces, technologies can have a positive impact on the psychosocial work environment, and therefore on workers' well-being. However, new technologies also have an impact on our personal life, the way we communicate and socialize. Electronic mail is considered as an indispensable communication means for most workers nowadays. Though it speeded up communications and thus increased organizational effectiveness, Hair *et al.* argued that e-mails are much more than the "electronification of the traditional mailed letter" (2007: 2791). Indeed, people tend to feel a need to respond quickly to e-mail, to check it while on vacation as well as throughout the day, and use it during work hours to deal with personal issues and vice versa (Czerwinski *et al.*, 2004; Jackson *et al.*, 2001, cited in Hair *et al.*, 2007). In the same way, the cell phone means people are always available to be contacted, whether it is for work- or nonwork-related purposes (Harrison *et al.*, 2004). Cell phones allow people to remain in closer contact with family and friends and they thus encourage social cohesion as people remain within familiar communal settings (Geser, 2004). Although e-mail or text messages can be perceived as cold and lacking humanity, some suggest that they provide more control over social interactions compared to synchronous communications like voice calls (Madell and Muncer, 2007).

On the other hand, ICT also has an impact on traditional physical barriers between work and personal life. By being accessible outside work hours or physical office locations, boundaries have been removed and customers, business partners, managers, and employees are always within reach no matter what time of the day or night (Kompier, 2006). Role boundary permeability "is the degree that a role allows one to be physically located in the role's domain, but psychologically or behaviourally involved in another role" (Ashforth *et al.*, 2000: 474). Technology use is often marketed to suggest that it promotes exactly this sort of role permeability and allows a more mobile workforce. Workers who have the technological means are mobile and do not have to work from traditional offices. Research findings are inconsistent, however, in terms of the consequences which this permeability has on workers' ability to maintain a balance between their work and personal life. Some authors argue that this flexibility allows them to meet both work and family needs more easily (Hill *et al.*, 2003). Others have raised serious concerns about work–life imbalance and work–family conflicts that such blurring of boundaries can cause (Hughes and Hans, 2001; Watt and White, 1999). For example, Chesley (2005) analyzed the consequences of persistent technology use (computer and communication technology) on individual well-being and family satisfaction. Her longitudinal study showed that persistent use of cell phones and pagers (not e-mails or Internet use) was significantly linked with distress and family dissatisfaction. The study in this volume by Kossek *et al.* also illustrates that the benefits of utilizing flexible technologies are not always apparent.

The blurring of boundaries between work and personal life due to ICT can also be associated with how we relate to time. Eriksen (2001) argued that, instead of time-saving, ICTs have actually made time more scarce, as responses to e-mails or text messages are expected to be instantaneous. Harrison and colleagues (2004) noted that this extension of acceptable work hours is often viewed unfavorably by workers:

> Within this new work/life culture, the balance of interest between employer and employee will need to be renegotiated. Through the last half of the 20th century, a consensus emerged on aspects of employment such as the length of the normal working week, with working time outside of these limits considered overtime. Expressions such as 9-to-5 reflect the widespread acceptance of this consensus, the idea of "24/7" potentially threatens this consensus. (Harrison *et al.*, 2004: 9)

Not only is this blurring of boundaries between private life and work life likely to be even more prevalent in the future, but the boundaries between what is "work" and leisure time are also likely to become more hazy (Costea *et al.*, 2005). Work, nonwork, and leisure are becoming harder to distinguish in a knowledge economy with high rates of technological uptake. An example of this blurring between work and leisure/personal time is Facebook, a networking web site, now being used by groups of various intellectual/business interests. In the United Kingdom, it has been reported that up to 70 percent of employers have actually banned access to Facebook during work hours, yet users claim they are building useful networks for the business (Frost, 2007). According to Sproull (2000), employees have traditionally used computers at work to meet personal and work-related interests, but Costea *et al.* (2005) argued that what is considered as work and what is play or personal time has and will continue to be increasingly blurred. A sales representative could be on a golf course or on Facebook networking and it could be considered as serious play . . . or playful work.

End-user reactions to technology

As noted earlier in this chapter, changes in ICT and related technologies have been found to exert substantial impact on the well-being of individual end-users. Here we discuss some of the major issues that have been explored in research. Our intention is not to provide an exhaustive review of all relevant issues, but rather to highlight some of the key variables that have been investigated and to provide some illustrations of the kinds of research conducted in this area. Hence this discussion will be indicative rather than attempting a comprehensive review of all research. We also refer readers to other recent valuable reviews, including Bessiere *et al.* (2006), Coovert *et al.* (2005), Olson and Olson (2003), and Thomee *et al.* (2007).

From a sociopsychological perspective, responses to technology can be grouped into three broad categories: cognitive (perceptions and beliefs about technology), affective (emotional and evaluative reactions), and behavioral (usage and withdrawal from use of technology). These three categories are clearly interrelated and interdependent, but most research attention has focused on end-user cognitive and affective (emotional) responses. As mentioned, positive outcomes can arise when individuals are able to utilize technology effectively and feel that it enables them to achieve important goals, including the development of self-efficacy and mastery, sense of goal achievement, and greater efficiency (Beas and Salanova, 2006). On the other hand, beliefs or perceptions of inadequacy (in operating the technology), failure to achieve goals, unmet performance targets, and so on can create significant psychosocial problems for individuals, both at work and in their "nonwork" environment (Batt and Valcour, 2003). Here we focus on three outcomes which have received considerable attention in research on technology usage – anxiety, frustration, and self-efficacy. We also briefly touch upon the potentially negative impact of ICT on social interactions in work environments.

Anxiety and depression

Numerous studies have explored the role played by (computer) anxiety and depression as critical affective reactions, although interestingly more attention has been accorded to anxiety than to depression. High levels of anxiety or depression can, of course, have a substantial impact on an individual's performance on any task or activity, but this is especially the case in respect of use of ICT. Anxiety may be caused by various factors, ranging from feelings of inadequacy (about one's ability to effectively utilize the technology) through to concerns about job insecurity (threat of job loss due to the introduction of *labor*-saving technology). In respect of job insecurity, Vieitez *et al.* (2001) observed a significant relationship between workers' beliefs that technological innovation would lead to job insecurity (and possibly loss of job opportunities) and their psychological well-being, including anxiety and depression. Workers in lower-status jobs and with less education were most fearful of the possibility of job loss due to new technology. Interestingly, however, there was no significant relationship between age and attitudes toward technology, which has been observed in other research on computer anxiety (see, e.g., Mikkelsen *et al.*, 2002).

The effects of anxiety are not constrained to job insecurity, however. At a more general level, anxiety about the actual use of ICT has been extensively investigated. Beckers and Schmidt (2001) developed and tested a six-factor model of computer anxiety which incorporated computer literacy, self-efficacy, physical arousal, affective reactions, and beliefs about both the benefits and dehumanizing effects of ICT. They found computer literacy had a strong positive influence on physical arousal and emotional (affective) responses, while self-efficacy contributed to increased literacy. Perhaps the most salient contribution of this study is its differentiation of various components of ICT anxiety, which enabled an assessment of specific relationships between computer usage and anxiety. In addition, the findings illustrated some very practical ways in which anxiety can be ameliorated.

Compeau and her colleagues (1999) longitudinally examined the influence of computer self-efficacy, outcome expectations, affect, and anxiety on computer usage in a large Canadian sample. As predicted, a significant negative relationship was found between anxiety and computer usage, while self-efficacy contributed to reduced anxiety. Contrary to expectation, however, no significant relationship was obtained between computer affect (e.g., liking, interest) and anxiety. The authors concluded that enhancement of self-efficacy is a key factor in the reduction of anxiety about the usage of ICT.

A similar investigation of ICT anxiety was reported recently by Czaja and her colleagues (2006), who tested a theoretical model of technological adoption (use of the WorldWideWeb) which included a range of predictor variables, among them computer ability, computer attitudes, and anxiety. Similar to Compeau *et al.*'s findings, Czaja *et al.* observed that computer efficacy played a central role in reducing computer anxiety. Although the latter was not strongly linked with overall technology usage, it was negatively associated with breadth of use, and anxiety mediated the effects of efficacy on usage. That is, more highly computer-anxious participants in this research reported a lower usage of this technology. Age was also a significant (negative) predictor of usage.

Another, relatively recent, dimension of ICT is use of the Internet, for both communication (with others) and information gathering. Internet utilization has escalated in recent years (Weiser, 2001), to the point where this technology has become a pervasive phenomenon in workplaces and outside of work (Joiner *et al.*, 2007). As with computer usage generally, however, anxiety about using the Internet can seriously limit the extent to which an individual is willing to explore this medium. For instance, in a Romanian study Durndell and Haag (2002) found that attitudes toward and anxiety about computer usage also extrapolated to Internet utilization – individuals who reported more anxiety were less likely to access the Internet. Similar results have been reported by Joiner *et al.* (2007) in a study of university students in the United Kingdom and Australia. While the majority of participants in this research were not anxious about using the Internet, those who were anxious were significantly less likely to utilize it.

A detailed discussion of anxiety in respect of ICT usage has been provided by Sami and Pangannaiah (2006), who distinguished three different manifestations of what they refer to as "technostress" or "technophobia," which may be viewed as an extreme symptom of psychological strain in relation to technology usage. The first manifestation they described as the "anxious technophobe," that is, a person who displays typical anxious physiological reactions (such as headaches, sweaty palms, increased heart rate) when using technology. The second kind is the "cognitive technophobe," who superficially appears to be relatively calm and relaxed about technology utilization, but internally is highly anxious about their capability and self-efficacy, as well as the likelihood of making mistakes. Finally there is the "uncomfortable user," an individual who may experience fewer symptoms of anxiety than the other categories, but who is nevertheless somewhat anxious and concerned about their ability to use ICT.

According to Sami and Pangannaiah, the above three types may exhibit somewhat different reactions to technology, although an underlying anxiety is present in all three. They noted that high levels of technophobia may result in avoidance *behaviors*, especially when combined with low confidence (self-efficacy) and negative attitudes toward technology. Based on the technology acceptance model developed earlier by Davis (1989), Sami and Pangannaiah suggested that anxiety will be associated with the perceived usefulness of the technology (that is, does the person believe it will help them to achieve their goals or to perform tasks more efficiently and effectively?) and the perceived ease of use. Individuals with lower expectations about the usefulness and ease of use of the technology are more likely to experience anxiety, as well as a tendency to withdraw from usage altogether.

It will be clear from the above discussion that anxiety about the use of technology can have severe consequences for an individual, in terms of both their work performance (productivity) as well as their psychological (and even physical) well-being. In addition, as documented by Coovert and Thompson (2003), prolonged anxiety can ultimately induce a state of exogenous depression, which may be very debilitating for an individual, in terms of both their job performance and their overall well-being. As with other sources of anxiety, concerns about one's ability to operate the technology to advantage, as well as insecurity about how to deal with problems which will inevitably arise, can lead the individual to develop an acute phobia which may ultimately translate into an inability to use the technology effectively, if at all. Such conditions can have a paralyzing effect on a person's motivation and capacity to utilize technology, which will inevitably engender a negative impact on their well-being (Coovert and Thompson, 2003).

Frustration

Alongside anxiety, another reaction that has commonly been observed in ICT usage, even among individuals who feel confident and efficacious in their use of this technology, is frustration. Most users experience momentary, and sometimes ongoing, feelings of frustration, which can have a deleterious effect on both their work performance and their psychological health and well-being. A comprehensive model of ICT frustration has been proposed by Bessiere and her colleagues (2006), who suggested that it can emerge in a variety of ways, although typically these are associated with an inability to perform tasks efficiently and to achieve important goals.

Bessiere *et al.* commented that while approximately 25 percent of people in the US do not have physical access to computers (or other ICT), around another 20 percent do not access this technology directly but use other people to send e-mails or access the Internet for them, and a further 17 percent have abandoned using the technology due to technical and related problems which engendered significant frustration. The computer frustration model outlined by Bessiere *et al.* offers a useful platform for understanding the processes through which frustration can emerge and the cognitive and emotional outcomes that can arise from continuing frustration with ICT usage.

Other researchers (e.g., Ceaparu *et al.*, 2004) have identified some of the major causes of frustration in respect of ICT usage. These include: unpredictable delays in program reaction times, poorly designed interfaces which may be difficult to utilize, time lost due to unclear error messages, excessively long downloading times, features being difficult to locate, and lost connections. Most ICT users will readily identify the experience of frustration resulting from the above (and other) difficulties, but reactions will of course depend on a range of other factors. One of these is the person's expectations. It has been demonstrated, for instance, that where an individual is "primed" to anticipate delays or error messages or long downloading time, their reactions will to some extent be mitigated by these expectations (Ceaparu *et al.*, 2004; Lazar *et al.*, 2006). Another potential problem commented upon by Coovert and Thompson, and also referred to by other commentators (see, e.g., Bessiere *et al.*, 2006; Ceaparu *et al.*, 2004), is that sophisticated technologies are not always easy to utilize and can create substantial frustration among users due to their complexity and sometimes unpredictability. The term "user-friendly" came into parlance as far back as 1977 (Coovert and Thompson, 2003), and referred to efforts to make technology more accessible and less daunting, especially for inexperienced and nontechnical users. Since then there have been continued efforts to design technological interfaces which better accommodate human capacity and needs, although not always successfully (Bessiere *et al.*, 2006).

A further issue that may contribute substantially to the individual's affective reactions is time pressure. Clearly, the need to perform tasks rapidly and to a deadline will create additional demands on the person, and will exacerbate their sense of frustration when problems occur with

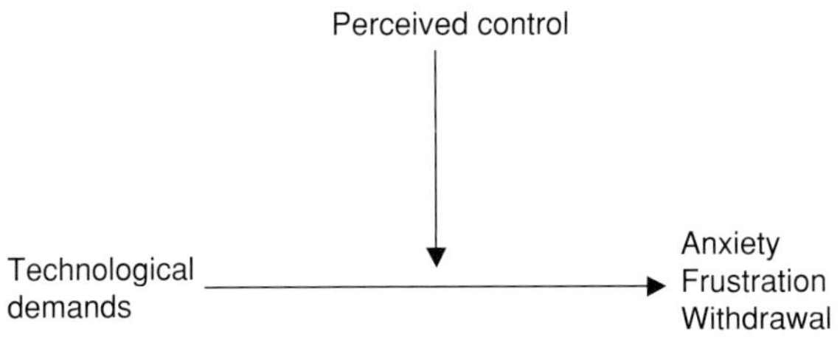

Figure 4.1 The moderating (buffering) effect of perceived control.

the technology. Organizations can assist in this regard, by offering appropriate (and tailored) training to workers utilizing ICT (Compeau *et al.*, 1999) – see later discussion of coping with technology.

The above overview suggests that certain factors may function as potential moderators (buffers) of negative reactions to ICT. These buffers may be classified as either personal (dispositional) or situational (environmental). Foremost among the personal or dispositional factors are variables such as attributions (internal versus external) and self-efficacy, which have already been referred to above. Below we briefly summarize the relevance of these variables to the acceptance and utilization of technology.

Attributions

Although the concept of locus of control (LOC) has been around for over forty years, only recently has it been invoked as a possible explanation of people's reactions to ICT and other technologies. LOC refers to a tendency to perceive events and outcomes in one's life as being under one's own control or as being controlled by sources over which the person has little or no control, such as luck, fate, or other people. In his landmark article on this topic, Rotter (1966) referred to these beliefs as internal and external LOC, and suggested that feelings of internal or external control are largely dispositional rather than situational. In this sense, LOC can be distinguished conceptually (and operationally) from the degree of control a person perceives oneself to have over specific environmental demands (such as workload). In practice, of course, the two processes tend to coexist (O'Driscoll, 2006).

Attributions about the degree of control the person feels they can exert over technology can be an important predictor of how they will cope with (a) new technological requirements, (b) changes in technology, and (c) difficulties encountered when using technology. The moderating or buffering effect is illustrated in Figure 4.1.

That is, when individuals are confronted by demands from the use of technology which may lead to symptoms of psychological strain (such as anxiety, frustration, and withdrawal), feeling that the situation is (at least to some extent) controllable will mitigate the negative impact of the stressors. On the other hand, when the person feels that they do not understand and cannot exert any influence over the difficulty (for example, they cannot determine how to resolve a failure or they do not know how to respond to an unexpected message), they will experience an increased level of anxiety and frustration, which may lead to withdrawal behavior.

This conceptualization of perceived control as a buffering variable accords with the Job Demand–Control model of stress proposed by Karasek (see Karasek and Theorell, 1990), and confirmed in subsequent research. Karasek suggested that it is not the level of demand per se which necessarily creates stress (or strain) for a person, but rather whether they feel that they can exert control over the demands which they encounter. Coovert and his colleagues (2003, 2005) have noted that perceived control can play an especially important role when new technology is being implemented within an organization. In these conditions, lack of perceived control over the technology is a major predictor of job dissatisfaction, and that "the lack of control experienced by workers who are required to deal with technological failures, inadequacies, and changes generate psychological stress and concomitant physical effects" (Coovert and Thompson, 2003: 229). Likewise, Beckers and Schmidt (2003) observed that feelings of control over computer technology were associated with more positive experiences (feeling computer literate) and with greater liking for ICT.

In a similar vein, Salanova *et al.* (2002) also found support for the Job Demand–Control model in a study of Spanish information technology workers. Individuals reporting higher levels of perceived control experienced fewer symptoms of burnout (emotional exhaustion and cynicism) when confronted with high demands than did their low-control colleagues. Put another way, when demands were high, levels of emotional exhaustion and cynicism were greater for workers feeling less autonomy in deciding how and when to perform their tasks. One practical implication of these findings is that increasing perceptions of control (over the work tasks and environment) may alleviate strain when it is difficult to modify job demands per se, and hence organizations should explore mechanisms for enhancing perceptions of internal (personal) control. Finally, as noted earlier, Kossek *et al.* (this volume) also demonstrate that feelings of control over work schedule flexibility are a critical determinant of the effects of teleworking on individuals' well-being.

Technological self-efficacy

A related, albeit distinct, concept to attributions of control is the individual's sense of self-efficacy, defined as "beliefs about one's capabilities to organize and execute the courses of action required to produce given attainments" (Bandura, 1997: 3). Various researchers have demonstrated that self-efficacy in relation to ICT has both direct (positive) effects and buffering effects on technology attitudes, satisfaction, and reduced anxiety, depression, and frustration. For example, in a study of communication technology workers, who might be anticipated to have *favorable* attitudes toward ICT, Beas and Salanova (2006) differentiated between three forms of self-efficacy beliefs: generalized, professional, and computer self-efficacy. Low levels of all three forms of self-efficacy were significantly associated with greater burnout, anxiety, and depression. Salanova *et al.* (2002) reported that the more specific the self-efficacy beliefs (that is, the more focused they were on technology per se), the greater their contribution to reduced burnout. Computer-related self-efficacy functioned to buffer the relationship between job demands and burnout levels, such that the impact of job demands on burnout was reduced when self-efficacy was high. Finally, Salanova *et al.* (2000) found that computer self-efficacy also moderated the relationship between computer training and burnout, suggesting that increasing self-efficacy *prior to* providing computer training may be beneficial in terms of reducing burnout.

Other researchers have obtained similar findings on the positive influence of self-efficacy. Bessiere *et al.* (2006), for instance, compared situational and dispositional predictors of technology frustration, finding that self-efficacy especially contributed to reduced frustration and more positive mood among computer users. In a longitudinal investigation of computer usage among Canadian subscribers to a business journal, Compeau *et al.* (1999) reported significant relationships between initial computer self-efficacy and outcome expectations, affect, anxiety, and usage one year later. The authors concluded that their findings "provide strong confirmation that both self-efficacy and outcome expectations impact on an individual's affective and *behavioural* reactions to information technology" (1999: 145).

Czaja and her colleagues (2006) examined the relationship between age and technology usage. Previous studies have suggested that older people have less favorable attitudes toward computer technology and are less likely to adopt it – the so-called "digital divide." Czaja *et al.* confirmed this relationship, but also found that the relationship between age and adoption of computer technology was moderated by computer self-efficacy (among other factors). Specifically, older people who were high in self-efficacy beliefs were more likely to adopt computer technology than their counterparts who were low in self-efficacy, whereas the difference was not so observable among younger people. Again, the beneficial effects of self-efficacy are illustrated by the findings of this research.

In sum, the studies discussed above (and others) suggest that feelings of mastery (self-efficacy) and perceived control over ICT can have a substantial influence on psychological health and well-being, in particular reducing anxiety (about using ICT) and depression (arising from failure to master the technology), and enhancing individuals' feelings of satisfaction with and intention to continue using this technology. These findings have implications for the design of interventions to increase individuals' skills and competency, as well as enhancing their overall well-being. As suggested by Salanova *et al.* (2000), early identification of levels of computer self-efficacy may be an important first step toward the development of effective interventions within organizations.

Individual coping and organizational strategies

Discussions of how to assist individual workers to effectively deal with the changes brought about by the introduction of new technology have typically been developed around (a) individual coping behaviors and (b) organizational strategies to minimize psychological (and in some cases physical) harm and to promote more effective usage of these technologies. The literature on stress-coping and stress-management processes is therefore very relevant to these considerations. In particular, the transactional model of stress-coping developed by Lazarus and his colleagues has frequently been the platform for discussions of adaptation to technological changes at the individual level.

Individual (or personal) coping is defined as "the cognitive and behavioural efforts exerted to manage specific external and/or internal demands that are appraised as taxing or exceeding the resources of the person" (Lazarus and Folkman, 1984: 141). From this perspective, at the heart of coping is the evaluation or appraisal that the person undertakes in relation to a demand (such as the requirement to adapt to a new ICT system at work). If the person perceives this demand as a "challenge," they will confront it in a very different manner than if they perceive it as a "threat" to their well-being. This distinction can make a very substantial difference in terms of the outcomes for an individual.

A detailed discussion of personal coping in relation to new technology has been provided by Beaudry and Pinsonneault (2005), who noted that a variety of theoretical models have now been developed to explain user adaptation processes and outcomes, but all of these have essentially derived from the Lazarus transactional perspective. As mentioned above, appraisal of the technology plays a critical role in the person's adaptation to it. For instance, if the person perceives that a new technology will make their job easier to perform, improve their performance, and create more satisfaction for them with their work, they are very likely to approach the innovation with enthusiasm and be motivated to overcome any obstacles to their mastery of it. Alternatively, if a technology has been foisted upon them and they cannot envisage any significant advantages for them personally, plus they have major concerns about their ability to learn to use the new technology effectively, chances are that they will perceive the change as a threat to their well-being.

Beaudry and Pinsonneault (2005) also noted that factors such as anxiety can contribute substantially to the manner in which an individual responds to technological change. Anticipatory anxiety may be reinforced by initial experiences in which the person feels that they cannot learn how to master the technology and where problems arise which they do not feel competent to resolve. This "cycle of anxiety can negatively influence beliefs about the technology and engender fear" (Beaudry and Pinsonneault, 2005: 499). Similarly, frustration over barriers to effective utilization of the technology (which we discussed earlier) can generate a sense of learned helplessness, blocking efforts to deal effectively with these challenges. In the terminology of the transactional model of stress-coping, individuals may then engage in *emotion-focused* coping (such as avoidance behavior) rather than *problem-focused* coping (making efforts to resolve the problems).

Beaudry and Pinsonneault (2005) outline four major user adaptation strategies: (1) maximizing benefits strategy, (2) satisficing benefits strategy, (3) disturbance-handling strategy, and (4) self-preservation strategy. The first strategy entails the person taking full advantage of the technology and hence maximizing their personal benefits (e.g., performance, satisfaction). In contrast, the second approach occurs when individuals feel that they have limited control over technology, even though they may recognize its advantages. Consequently, limited adaptation to the technology occurs. A disturbance-handling strategy is utilized when the person perceives IT as a threat (to their well-being), although they also believe they can exert control, and therefore they make efforts to reduce the negative impact of IT on their well-being. Finally, self-preservation occurs when IT is perceived as a threat and the person feels little or no control over it. Under these conditions, adaptation efforts will be predominantly focused on changing their emotional response to IT rather than increasing their ability to utilize it. In a sense, these four strategies can be viewed along a continuum of problem-focused to emotion-focused coping. Clearly, the ideal would be for end-users to adopt a maximizing benefits strategy. For this to occur, they need to feel a strong sense of opportunity to utilize the technology and address difficulties, coupled with a high level of perceived control (see our discussion above of this variable). In reality, for many end-users there can be an oscillation between strategies 2, 3, and 4, and in cases where individuals experience both little opportunity and low control, self-preservation is more likely to occur. One potential outcome of this coping response could be withdrawal from using the technology, and a subsequent perpetuation of the belief that the person cannot master it, which may have negative consequences for their self-concept and general feelings of well-being.

In sum, one important consideration in addressing user adaptation is to explore the individual's appraisals of the technology, to ascertain whether they perceive it as a challenge versus a threat, and to determine their likely coping strategy. This should not be just an individual responsibility, but one which is shared by the person and the organization. Various authors have discussed what organizations might do to assist their workers to adapt more effectively to ICT and other technologies, in particular to changes in technology, which can have a substantial impact on worker performance (Beas and Salanova, 2006; Coovert and Thompson, 2003; Smith *et al.*, 1999; Smith and Carayon, 1995). The most obvious assistance that organizations can provide is training (in how to use the technology and how to attend to problems which arise). However, the general literature on training suggests that training programs and efforts do not always result in effective learning and behavior change, for several reasons.

First, the training which is offered may not be attuned to the needs and capacities of individual end-users. For example, older workers may have less facility to learn computer-based skills (Czaja *et al.*, 2006), may require a slower-paced learning environment, and may experience more anxiety and concerns about using ICT to undertake tasks that are either completely new to them or which they had previously performed in some other manner (e.g., manually). Furthermore, training can have different outcomes for different people and it is critical to consider individual differences (e.g., in learning styles and preferences) when implementing training, and to examine factors which may moderate the relationship between technological demands and people's cognitive and affective experiences (Bessiere *et al.*, 2006).

The experience of anxiety, frustration, and lack of control over the technology, coupled with how individuals endeavor to cope, both with the technology itself and with their affective reactions to it, may have a very substantial impact on their ability to utilize it effectively and their subsequent psychological well-being (Beckers and Schmidt, 2003; Joiner *et al.*, 2007). In brief, training initiatives must be tailored to accommodate individual needs and competencies, rather than adopting a one-size-fits-all approach. Given the accelerating rate of technological change (Coovert *et al.*, 2005; Hughes and Hans, 2001; Mamaghani, 2006), the specific knowledge and competencies needed to effectively engage with current technologies may quickly become outdated and superseded by new requirements, hence it is vital that individual workers have the opportunity to develop competencies that are generalizable and transferable to other contexts. Developing a sense of mastery and control is essential to the maintenance of psychological health and well-being (Beas and Salanova, 2006; Beaudry and Pinsonneault, 2005). It follows that training activities need to be geared to enhancing individuals' sense of mastery over new technology generally, as well as the specific skills needed to operate a particular technology.

Finally, although much has been written about the need for appropriate kinds and levels of support to be provided to end-users, the implementation of effective support mechanisms still appears to be infrequent and haphazard. There would appear to be insufficient cross-fertilization between technology implementation and the literature of stress-management, especially the role of social support in enabling individuals to develop more effective coping strategies. Anecdotal reports suggest that individuals frequently experience frustration with the kinds of support which are provided by ICT specialists, indicating that there is a gap between the needs of end-users and the capacity of specialists to fulfill those needs. There is, however, a need for more empirical research on the most effective strategies for providing end-user support, including the kinds of support required, the timing of support, and its frequency.

Conclusion and practical recommendations

In this chapter we have endeavored to summarize some key issues concerning the relationship between the use of technology (especially ICT) and individual psychological well-being. As noted at the outset, we have not attempted to provide an exhaustive review of the vast array of research which has been conducted in this field, but rather to highlight a few salient issues that have emerged from this productive line of research as well as some of the impacts technology has had on the psychosocial work environment, and identify some of the major variables affecting individuals' reactions and well-being. From this admittedly selective overview we suggest some key considerations for organizations as they introduce and implement new information and communication technologies. To date, relatively scarce attention has been given to the psychosocial dimensions of technological change (Mamaghani, 2006), largely because the main thrust has been to advance the technology per se rather than to consider its impact on individual workers. As social scientists, however, we have the knowledge and mandate to bring a social and psychological perspective to new technology development and implementation.

As outlined in this chapter, research has demonstrated that technological changes are closely related to the psychosocial work environment and have a major influence on job demands, job control, and social relationships in the workplace. Given the criticality of these variables for workers' stress, interventions designed to improve the psychosocial work environment should consider how ICT can be used as a means toward the goal of reducing stress. On the other hand, technological changes have the potential to increase workers' stress, especially when not enough consideration is paid to the development of feelings of mastery (or control) and self-efficacy. The above discussion illustrates that considerably more attention must be accorded to the enhancement of these (and other) psychosocial factors. The absence of mastery and self-efficacy has been shown in numerous studies to be associated with higher anxiety, increased frustration, and ultimately a greater probability of exogenous depression. The provision of training, learning experiences, and support programs which take these factors into account will be essential to enhance individual worker well-being, as well as optimizing performance levels. As noted, this training, learning, and support needs to be tailored to the specific requirements and capabilities of individuals within the organization, as well as ensuring that competencies developed are transferable to other settings.

Finally, while the notion of "user-friendly" technology has been in existence now for many years (Coovert and Thompson, 2003), technological interfaces are still often designed with little acknowledgement of how individuals function in reality (Bessiere *et al.*, 2006; Ceaparu *et al.*, 2004), in particular how the technological demands will impinge upon the person's coping abilities and limitations. The design, development, and implementation of technology (especially ICT) require a thorough understanding of how people think, feel, and respond when dealing with systems that challenge their capacity and may not be fully under their control.

References

Armstrong, D., and Cole, P. (1995). Managing distances and differences in geographically distributed work groups. In S. Jackson and M. Ruderman (Eds.), *Diversity in Work Teams: Research paradigms for a changing workplace.* Washington, DC: American Psychological Association.

Ashforth, B. E., Kreiner, G. E., and Fugate, M. (2000). All in a day's work: Boundaries and micro role transitions. *Academy of Management Review*, 25, 472–491.

Bandura, A. (1997). *Self-efficacy: The exercise of control*. New York: Freeman.

Batt, R., and Valcour, P. M. (2003). Human resources practices as predictors of work–family outcomes and employee turnover. *Industrial Relations*, 42(2), 189–220.

Beas, M. I., and Salanova, M. (2006). Self-efficacy beliefs, computer training and psychological well-being among information and communication technology workers. *Computers in Human Behavior*, 22(6), 1043–1058.

Beaudry, A., and Pinsonneault, A. (2005). Understanding user responses to Information Technology: A coping model of user adaptation. *MIS Quarterly*, 29(3), 493–524.

Beckers, J. J., and Schmidt, H. G. (2001). The structure of computer anxiety: A six-factor model. *Computers in Human Behavior*, 17(1), 35–49.

(2003). Computer experience and computer anxiety. *Computers in Human Behavior*, 19(6), 785–797.

Bessiere, K., Newhagen, J. E., Robinson, J. P., and Shneiderman, B. (2006). A model for computer frustration: The role of instrumental and dispositional factors on incident, session, and post-session frustration and mood. *Computers in Human Behavior*, 22(6), 941–961.

Bliese, P. D., and Britt, T. W. (2001). Social support, group consensus, and stressor–strain relationships: Social context matters. *Journal of Organizational Behavior Management*, 22(4), 425–436.

Bond, F. W., and Bunce, D. (2001). Job control mediates change in a work reorganization intervention for stress reduction. *Journal of Occupational Health Psychology*, 6(4), 290–302.

Carayon-Sainfort, P. (1992). The use of computers in offices: Impact on task characteristics and worker stress. *International Journal of Human–Computer Interaction*, 4(3), 245.

Ceaparu, I., Lazar, J., Bessiere, K., Robinson, J., and Shneiderman, B. (2004). Determining causes and severity of end-user frustration. *International Journal of Human–Computer Interaction*, 17(3), 333–356.

Chesley, N. (2005). Blurring boundaries? Linking technology use, spillover, individual distress, and family satisfaction. *Journal of Marriage and Family*, 67(5), 1237–1248.

Clegg, C. (1994). Psychology and information technology: The study of cognition in organizations. *British Journal of Psychology*, 85(4), 449.

Compeau, D., Higgins, C. A., and Huff, S. (1999). Social cognitive theory and individual reactions to computing technology: A longitudinal study. *MIS Quarterly*, 23(2), 145.

Coovert, M. D., and Thompson, L. F. (2003). Technology and workplace health. In J. C. Quick and L. E. Tetrick (Eds.), *Handbook of Occupational Health Psychology* (pp. 221–241). Washington, DC: American Psychological Association.

Coovert, M. D., Thompson, L. F., and Craiger, J. P. (2005). Technology. In J. Barling, E. K. Kelloway, and M. R. Frone (Eds.), *Handbook of Work Stress* (pp. 299–324). Thousand Oaks, CA: Sage Publications.

Costea, B., Crump, N., and Holm, J. (2005). Dionysus at work? The ethos of play and the ethos of management. *Culture and Organization*, 11(2), 139–151.

Czaja, S. J., Charness, N., Fisk, A. D., Hertzog, C., Nair, S. N., Rogers, W. A., and Sharit, J. (2006). Factors predicting the use of technology: Findings from the Center for Research and Education on Aging and Technology Enhancement (CREATE). *Psychology and Aging*, 21(2), 333.

Czerwinski, M., Horvitz, E., and Wilhite, S. (2004). A diary study of task switching and interruptions. Paper presented at the SIGCHI conference on human factors in computing systems, April 24–29, New York, NY.

Davis, F. D. (1989). Perceived usefulness, perceived case of use and user acceptance of information technology. *MIS Quarterly*, 3, 319–340.

Durndell, A., and Haag, Z. (2002). Computer self efficacy, computer anxiety, attitudes towards the Internet and reported experience with the Internet, by gender, in an East European sample. *Computers in Human Behavior*, 18(5), 521–535.

Eriksen, T. H. (2001). *Tyranny of the Moment: Fast and slow time in the information age*. London: Pluto Press.

Frost, V. (2007). Contact sports. *Guardian*, Aug. 25, 1–2.

Geser, H. (2004). Towards a sociological theory of the mobile phone 3.0. Retrieved August 10, 2007, from http://socio.ch/mobile/t_geser1.pdf.

Gustafsson, E., Dellve, L., Edlund, M., and Hagberg, M. (2003). The use of information technology among young adults – Experience, attitudes and health benefits. *Applied Ergonomics*, 34(6), 565.

Hair, M., Renaud, K. V., and Ramsay, J. (2007). The influence of self-esteem and locus of control on perceived email-related stress. *Computers in Human Behavior*, 23(6), 2791–2803.

Hamborg, K. C., and Greif, S. (2003). New technologies and stress. In M. J. Schabracq, J. A. M. Winnubst, and C. L. Cooper (Eds.), *The Handbook of Work and Health Psychology*, 2nd edn. Chichester: John Wiley & Sons Ltd.

Harrison, A., Wheeler, P., and Whitehead, C. (2004). *The Distributed Workplace*. London: Spon Press.

Hill, E. J., Ferris, M., and Martinson, V. (2003). Does it matter where you work? A comparison of how three work venues (traditional office, virtual office, and home office) influence aspects of work and personal/family life. *Journal of Vocational Behavior*, 63, 220–241.

Hughes, R., and Hans, J. D. (2001). Computers, the Internet, and families. *Journal of Family Issues*, 22(6), 776–790.

International Telecommunication Union. (2008). Worldwide mobile cellular subscribers to reach 4 billion mark late 2008 [Electronic Version]. *ITU Magazine*. Retrieved February 15, 2008, from www.itu.int/newsroom/press_releases/2008/29.html.

Introna, L. D. (2001). Cooperation, coordination and interpretation in virtual environments: Some thoughts on working-together. *Cognition, Technology & Work*, 3(2), 101–110.

Jackson, T., Dawson, R., and Wilson, D. (2001). The cost of email interruption. *Journal of Systems and Information Technology*, 5, 81–92.

Johnson, J. V., and Hall, E. M. (1988). Job strain, workplace social support, and cardiovascular disease: A cross-sectional study of a random sample of the Swedish working population. *American Journal of Public Health*, 78(10), 1336–1342.

Johnson, J. V., Hall, E. M., and Theorell, T. (1989). Combined effects of job strain and social isolation on cardiovascular disease, morbidity and mortality in a random sample of the Swedish male working population. *Scandinavian Journal of Work, Environment & Health*, 15, 271–279.

Joiner, R., Brosnan, M., Duffield, J., Gavin, J., and Maras, P. (2007). The relationship between Internet identification, Internet anxiety and Internet use. *Computers in Human Behavior*, 23(3), 1408–1420.

Jordan, J., Gurr, E., Tinline, G., Giga, S., Faragher, B., and Cooper, C. (2003). *Beacons of Excellence in Stress Prevention*, Research report 133. Manchester: Health and Safety Executive.

Karasek, R. (1979). Job demand, job decision latitude, and mental strain: Implications for job redesign. *Administrative Science Quarterly*, 24(2), 285–307.

Karasek, R. A., and Theorell, T. (1990). *Healthy Work*. New York: Basic Books.

Kendall, J. E., and Kendall, K. E. (1993). Metaphors and methodologies: Living beyond the systems machine. *MIS Quarterly*, 17, 149–171.

Khor, Z., and Marsh, P. (2006). Life online: The Web in 2020. Retrieved August 4, 2007, from www.scribd.com/doc/302121/Life-Online-The-Web-in-2020.

Kompier, M. A. J. (2006). New systems of work organization and workers' health. *Scandinavian Journal of Work, Environment & Health*, 32(6; special issue), 421–430.

Korunka, C., Weiss, A., Huemer, K.-H., and Karetta, B. (1993). Effects of new technologies with special regard for the implementation process per se. *Journal of Organizational Behavior*, 14, 331–348.

Kostakos, V., O'Neill, E., Little, L., and Sillence, E. (2005). The social implications of emerging technologies. *Interacting with Computers*, 17(5), 475–483.

Lazar, J., Jones, A., and Shneiderman, B. (2006). Workplace user frustration with computers: An exploratory investigation of the causes and severity. *Behaviour & Information Technology*, 25(3), 239–251.

Lazarus, R. S., and Folkman, S. (1984). *Stress Appraisal and Coping*. New York: Springer.

Madell, D. E., and Muncer, S. J. (2007). Control over social interactions: An important reason for young people's use of the Internet and mobile phones for communication? *CyberPsychology & Behavior*, 10(1), 137–140.

Mamaghani, F. (2006). Impact of information technology on the workforce of the future: An analysis. *International Journal of Management*, 23(4), 845.

Marshall, G. W., Michaels, C. E., and Mulki, J. P. (2007). Workplace isolation: Exploring the construct and its measurements. *Psychology and Marketing*, 24(3), 195–223.

Melchior, M., Niedhammer, I., Berkman, L. F., and Goldberg, M. (2003). Do psychosocial work factors and social relations exert independent effects on sickness absence? A six year prospective study of the GAZEL cohort. *Journal of Epidemiology and Community Health*, 57(4), 285–293.

Mikkelsen, A., Ogaard, T., Lindoe, P. H., and Olsen, O. E. (2002). Job characteristics and computer anxiety in the production industry. *Computers in Human Behavior*, 18, 223–239.

Nielsen, K., Randall, R., and Albertsen, K. (2007). Participants, appraisals of process issues and the effects of stress management interventions. *Journal of Organizational Behavior*, 28, 793–810.

O'Driscoll, M. P. (2006). Locus of control. In J. H. Greenhaus and G. A. Callanan (Eds.), *Encyclopedia of Career Development* (pp. 480–482). Thousand Oaks, CA: Sage Publications.

Olson, G. M., and Olson, J. S. (2003). Human–computer interaction: Psychological aspects of the human use of computing. *Annual Review of Psychology*, 54(1), 491–516.

Rotter, J. B. (1966). Generalized expectancies for internal versus external control of reinforcement. *Psychological Monographs*, 80, 1–28.

Salanova, M., Grau, R. M., Cifre, E., and Llorens, S. (2000). Computer training, frequency of usage and burnout: The moderating role of computer self-efficacy. *Computers in Human Behavior*, 16(6), 575–590.

Salanova, M., Peiro, J. M., and Schaufeli, W. B. (2002). Self-efficacy, specificity and burnout among information technology workers: An extension of the job demand-control model. *European Journal of Work and Organizational Psychology*, 11(1), 1–25.

Sami, L. K., and Pangannaiah, N. B. (2006). "Technostress": A literature survey on the effect of information technology on library users. *Library Review*, 55(7), 429–439.

Smith, M. J., and Carayon, P. (1995). New technology, automation, and work organization: Stress problems and improved technology implementation strategies. *International Journal of Human Factors in Manufacturing*, 5(1), 99–116.

Smith, M. J., Conway, F. T., and Karsh, B. T. (1999). Occupational stress in human–computer interaction. *Industrial Health*, 37(2), 157–173.

Sproull, L. S. (2000). Computers in U.S. households since 1977. In A. D. J. Chandler and J. W. Cortada (Eds.), *A Nation Transformed by Information* (pp. 257–280). London: Oxford University Press.

Thomee, S., Eklof, M., Gustafsson, E., Nilsson, R., and Hagberg, M. (2007). Prevalence of perceived stress, symptoms of depression and sleep disturbances in relation to information and communication technology (ICT) use among young adults – An explorative prospective study. *Computers in Human Behavior*, 23(3), 1300–1321.

Vieitez, J. C., Carcia, A. D., and Rodriguez, M. T. V. (2001). Perception of job security in a process of technological change: Its influence on psychological well-being. *Behaviour & Information Technology*, 20(3), 213–223.

Wastell, D. G., and Newman, M. (1996). Stress, control and computer system design: A psychophysiological field study. *Behaviour & Information Technology*, 15(3), 183–192.

Watt, D., and White, J. M. (1999). Computers and family life: A family development perspective. *Journal of Comparative Family Studies*, 30, 1–15.

Weiser, E. B. (2001). The functions of Internet use and their social and psychological consequences. *CyberPsychology & Behavior*, 4(6), 723–743.

5 From ergonomics to hedonomics: trends in human factors and technology

Tal Oron-Gilad and Peter A. Hancock

Ergonomics (or human factors) has been defined as *the application of scientific information concerning objects, systems, and environment for human use* (International Ergonomics Association, 2007). Ergonomics is commonly thought of in terms of how companies design tasks and work areas to maximize the efficiency and quality of their employees' work. However, ergonomics comes into everything which involves people and technology and largely concerns the physical and cognitive interactions between people and their respective creations. Helander (1997) specified three important targets for ergonomics design activity: to improve (1) safety; (2) productivity; and (3) operator satisfaction. It is important to note that the targets for ergonomics design are constantly evolving and what Helander stated now more than a decade ago may already seem outdated to many, yet unattainable to others. To understand the changes that are rapidly occurring in this evolving area of science, it is important, first, to briefly look back at the tremendous progress and change of focus that occurred in work and work design over the past century and especially recent decades.

Trends in ergonomics

Prior to the Industrial Revolution, it is the work of two individuals, Ramazzinni and Jastrzebowski, that is almost alone in being solely directed to the study of work design and ergonomics. Bernardino Ramazzinni was an Italian physician (1633–1714) who, based on his medical practice and experience, wrote what is considered to be the first report on work-related complaints. Ramazzinni's (1713) treatise became especially relevant when, in 2001, the US Congress voted to undo the nation's ergonomics standard, passed several months earlier in 2000. The repealed Occupational Safety and Health Administration (OSHA) standard addressed common occupational threats. The OSHA standard would have required employers to create programs to inform workers about musculoskeletal disorders and their various symptoms; to encourage reporting; and, if a serious incident were to occur, to evaluate the workplace and take action to address any risks identified (and see Parasuraman, Hancock, Radwa, and Marras, 2003). The standard received significant opposition from business groups, who claimed that it would be too costly to enact (Fee and Brown, 2001). These later authors indicated that publishing excerpts from Ramazzinni's writings would serve to remind us as a society that the battle over prevention of occupational hazards is far from being over, even though such occupational hazards have been documented by now for over three centuries.

Wojciech Jastrzebowski, a Polish biologist, coined the term ergonomics in 1857 in a philosophical treatise which was "based upon the truths drawn from the Science of Nature" (Jastrzebowski, 1857). In his original description of ergonomics, Jastrzebowski intended "work" to have a very broad meaning, which still has not lost its relevance today: "[T]his Science of Work, understood as Work in the comprehensive and integral sense, not merely its part that is physical labour or toil, but physical, aesthetic, rational, and moral work, that is Labour, Entertainment, Reasoning, and Dedication" (1857). His broad definition of ergonomics incorporated ideas related to both worker well-being and satisfaction. It is a much broader definition than simply the fundamental and morally laudable concept of the prevention of pain and occupational hazards.

No significant progress in the science of ergonomics was evident until the early 1900s (Meister, 1999). Frederick Taylor and his scientific management approach (Taylor, 1911) represented the next evolutionary step in the development of modern work systems. Taylor was the first to provide insight and proof of the simple fact that increasing individual workers' productivity could result in giving employment to more workers, rather than make others redundant. Taylor focused on efficiency. He recognized that proper compensation (i.e., additional wages that the workmen were able to earn beyond what they received for a daily payment) coupled with daily close contact with the management as well as training workers to follow predefined processes (i.e., a documented process for each work element rather than encouraging common rule-of-thumb methods) eliminated soldiering among workers and increased productivity. Taylor suggested it was important to "select and then train, teach, and develop the workman, whereas in the past he chose his own work and trained himself as best he could" (1911: 5–29). He also defined the role of management as being equally responsible for the success of the process as the worker himself: "There is an almost equal division of the work and the responsibility between the management and the workmen. The management must take over all work for which they are better fitted than the workmen, while in the past almost all of the work and the greater part of the responsibility were thrown upon the men" (1911: 5–29).

Early production environments were largely dependent on human power and human motion. The Scientific Management approach improved worker efficiency by specifying the workflow and the use of tools in the production process. Taylor sought the "One Best Way" in which tasks could be performed. In one well-known example at Bethlehem Steel, Taylor dramatically increased worker production and wages in a shoveling task by matching the shovel with the type of material that was to be moved, that is, ashes, coal, or ore. This approach could be referred to as one of the earliest attempts to develop a use-centered design. *Use-centered design* is a design construct that demands an integration of user, instrument, and goal. Thus, distinctions between user, instrument, and goal are sublimated and new distinctions emerge, in contrast with the more well known concept of *user-centered design* that differentiates the human factor as a special, distinct component within the system. (See Flach and Dominguez, 1995, for a more contemporary discussion of this former concept.) Henry Gantt, one of Taylor's colleagues and followers, developed the commonly used Gantt chart. His essays *Work, Wages, and Profits* (1913) and *Organizing for Work* (1919) argued for daily organization and planning of the employees' workday by defining tasks and their order of completion. Taylor's followers, Frank and Lillian Gilbreth, made jobs more efficient and less fatiguing through time and motion analysis, and standardizing tools, materials, as well as the job process. One of their famous applications was in bricklaying: the number of motions in bricklaying was reduced from 18 to 4.5, allowing bricklayers to increase their pace of laying bricks from 120 to 350 per hour. Another contribution by Lillian Gilbreth (1956) was related to the design of efficient household interiors (e.g., kitchens), household appliances (e.g., kitchen mixer and the interior setup of the refrigerator), and household task-based work processes. Gilbreth's designs are probably the first attempts to devise ergonomically efficient and safe environments for the home. Graham (1997), among others, has argued that Gilbreth's contribution to home management actually had a negative influence on the modern, urban, middle-class American woman, particularly with regard to promoting consumption and setting standards of quality that were not necessarily based on needs, but rather on desires (i.e., promoting the use of a kitchen mixer and/or preaching the consumption of advanced technological appliances for the kitchen does not necessarily improve the quality of the food made in that kitchen). We return to the question of efficiency and ergonomics design in the home later in this chapter.

In the industrial realm, the broad and rather positive acceptance of Taylorism in industry and manufacturing, in particular, reinforced a simple principle: improving workers' well-being is almost always tightly coupled with organizational profit. Hence, workers' conditions can only be improved when their employers foresee the immediate benefit to the workplace from the process. Unfortunately, Taylorism also had its downsides; those were mainly related to the loss of skill, loss of motivation, and the proliferation of dull and unfulfilling jobs. As the use of machines increased, workers were required to do the tasks that machines could not perform (e.g., move carts from one machine to the other, pack and stack products, cleaning). Although machines were developed to do the Dull, Dutiful, and Dirty tasks, under this compulsion, human operators began to share the same destiny. Taylor's principles were often misinterpreted and abused. Manufacturers wanted to reduce training cost and training time in order to be able to replace workers "on the fly" and they focused on generating work environments where very little job training and skill was required.

Eventually, as the need to control, maintain, and monitor machines increased, new types of jobs emerged. These emergent tasks required skilled and trained people. Thus, the increased use of technology generated contradictory routes and variations in jobs: (a) the removal/separation of the human operator from the physical properties of the action and the immediate feedback loop of the work process caused people to lose their skills in respect of the process itself; therefore, many workers lost their traditional trade skills, and jobs became dull and unfulfilling; (b) new types of jobs emerged: for example, monitoring, controlling, and maintaining machines, which required more highly trained operators; and (c) technology was enhanced by the human ability to explore places, distances, and tasks as never before, which further generated another type of job for the most highly trained and select people (e.g., pilots). This distinction between classes of workers still remains current and is again influential with regard to how important workers' well-being is to their employers.

The Second World War led to an exponential growth in technological development. Among these advances were sophisticated aircraft, radar, and visual information systems, sonar systems, as well as complex information integration. This prompted greater interest in human–machine interaction, mainly in the military, whose primary objective is to effect action at a distance as safely and efficiently as possible (see Hancock and Szalma, 2008). The efficiency of sophisticated military equipment was on multiple occasions directly compromised by bad or confusing interface design. Design concepts for "fitting the machine to the soldier" and usable control panels evolved (e.g., US Army MANPRINT [man power and personnel integration program]). Advanced selection and training of individuals to operate these complex systems became ever more necessary. It became evident that operator–machine interfaces were the medium through which humans affect the world. With regard to this era, 1949 is often cited as the inception benchmark of ergonomics as an official multidisciplinary science involving psychologists, physiologists, cognitive engineers, engineers, and occupational therapists (see also Craik, 1948). This is also when the use of the concept "usability" started (although the term itself is much more associated with Nielsen and his usability engineering book [1993]).

Gradually, usability increased and over the years has become formalized and defined as seeking three goals: effectiveness, efficiency, and user satisfaction (Preece *et al.*, 2002). *Effectiveness* is the accuracy and completeness with which users achieve certain goals (Frøkjær *et al.*, 2000). Indicators of effectiveness include quality of solution and error rates. *Efficiency* is the relation between (1) the accuracy and completeness with which users achieve certain goals and (2) the resources expended in achieving them. Indicators of efficiency include task completion time and learning time (Frøkjær *et al.*, 2000). *User satisfaction* is the users' comfort with and positive attitudes toward the use of the system. User satisfaction can be measured by subjective ratings or psychophysiological measurements. Of these three goals, the first two have received the most attention. Satisfaction was rarely measured as anything more than user preference. Contemporary technical developments have vastly magnified this opportunity for distal interaction through the medium of automated and semiautomated technologies. Advances in robot technology now open tremendous vistas of opportunity, but interaction with the human controller remains the crucial bottleneck for effectiveness of even such highly automated technical innovations. Action frequently provides real-time response feedback, letting the person know the outcome of his/her efforts in order to match outcome to intention. Dissociation at this juncture is problematic since the act of perception is an active one and divorcing the process of sensory assimilation from perceptual interpretation promises to fracture one of the fundamental human capacities (Gibson, 1966, 1979; Powers, 1973). Little wonder then that "situation awareness" represents the crucial bottleneck in current systems (Endsley, 1995; Smith and Hancock, 1995). It is evident that human performance is vastly disrupted if this feedback loop is distorted in either the spatial or the temporal dimensions. What remains unknown is how to frame such information exchange to best facilitate human decision and action response within such a general remote perception–action system (see Hancock *et al.*, 1995).

Gradually, the success of technological achievements penetrated into the more obvious commercial realms of everyday life (e.g., automobiles). The accessibility of sophisticated technology to the general public meant less selection or indeed no selection of users at all. This required the development of new ergonomic metrics for adaptation to operator characteristics. Designers are now required to address the large variability among users and consider elements that were traditionally covered by training or selection. Addressing user characteristics, traits (style and personality), motives (i.e., intrinsic and extrinsic motivation), skills, and goals for life (personal motives, beliefs, and worldview) is now part of the design process (see, e.g., Chapanis, 1988). The increased use of personal computers, combined with improved communication technologies and the WorldWideWeb, contributed to the erasure of the distinction between working at home and working at specific workplaces, as well as of the distinctions between tasks performed during "leisure" as compared with tasks performed during "work." It is in this realm that the concept of hedonomics has now emerged.

The current millennium

As a companion concept to ergonomics, hedonomics is a related area of investigation defined as "that branch of science which facilitates the pleasant or enjoyable aspects of human–technology interaction" (Hancock *et al.*, 2005). It has two main purposes: the promotion of pleasant and enjoyable human–system interaction, and the promotion of well-being and use of technology to facilitate such well-being. Thus, hedonomics is an allied endeavor directed primarily at the affective dimension of the interaction between people and the environment and artifacts they create. The term hedonomics is derived from two Greek roots: *hedon(e)*, meaning joy or pleasure, and *nomos*, meaning law-like or collective. In contrast to ergonomics, whose moral ethos is founded in efforts to prevent pain, injury (mostly repetitive, work-related), and suffering (and even fatality) predominately in the workplace, hedonomics is much more directed at the positive aspects of work interaction. Ergonomics is now acknowledged as the formal branch of science devoted to improvements in the physical and cognitive environments, serving to enhance the quality of life (see, e.g., *The Bathroom* by Alexander Kira [1966] as a classic example of ergonomic-oriented design in the home). In the same realm of human–technology interaction, hedonomics is primarily concerned with the promotion of pleasure. The major difference then between hedonomics and human factors or ergonomics (HF/E) is that HF/E is often handicapped by trying to show the value of preventing events that eventually do not happen or we want to avoid happening. In contrast hedonomics, being a positive enterprise, is intended to promote the occurrence of a pleasurable/satisfying event and we can actually demonstrate this positive outcome.

The history of hedonomics is relatively short and it appeared in various forms around the beginning of the current millennium. Some of the pioneering ideas/terms were funology (Blythe *et al.*, 2003), design for pleasure (Jordan, 2000), and affective and pleasurable design (Helander and Khalid, 2005; Khalid, 2004). Given that hedonomics is a hybrid, practical, and interdisciplinary area, the base knowledge is drawn from multiple fields both in the sciences and the humanities. Hedonomics draws heavily on discipline-based constructs from positive psychology: emotion, motivation, and affect (Snyder and Lopez, 2002). However, the expectation is that this form of science will develop into an individual and distinct area of study in and of itself. Affective evaluations provide new and different perspectives in human factors engineering. It is not a question of how to evaluate users – it is how the user evaluates. The research on hedonic values and seductive interfaces is in fact a welcome contrast to safety and productivity, which have traditionally dominated human factors and ergonomics (see Murphy *et al.*, 2003). Consequently, emotions and affect have received increasing attention over recent years. Approaches to emotions and affect have been studied on many different levels, and several models have been proposed for a variety of domains and environments.

Hedonomics, as it develops, is likely to draw on theories of motivation and emotion for advancing the goal of facilitating the development of convivial (Illich, 1973) and pleasurable technologies. Although there is an overabundance of theories of motivation and emotion, general themes emerge that are relevant for a mature hedonomic science. Thus, most theories incorporate goals (e.g., Locke and Latham, 1990), often organized into hierarchies, as a central driver of motivation and emotional states. Relevant issues include how goals are selected, the consequences for meeting or failing to meet those goals, and the rate of progress toward goal states. Another central theme is that of appraisal, in which individuals' motivational and emotional states are determined by their subjective evaluation of events (Lazarus, 1991; Scherer, 1999). These appraisals are influenced by estimates individuals make regarding the relative likelihood of outcomes (expectancies) and the personal relevance of those outcomes (valences and values; e.g., George and Jones, 1997; Vroom, 1964). For hedonomic interventions to succeed, they must facilitate genuine fulfillment of these needs. From this perspective, a simple usability approach is likely to fail, since it is based upon user preferences rather than whether the technology facilitates the individual's needs for competence, autonomy, and relatedness. Further, for hedonomic design to reach its fullest potential in work environments, the organizational/social context must be considered as part of the system to be designed. This will require a comprehensive sociotechnical approach incorporated into the design process.

As in other areas of human factors, purpose will be the central issue in the application of motivation and emotion theory to hedonomic design (Hancock, 1997; 2009). How one applies these theoretical models (and which theories are adopted) and which hedonomic goals are selected will depend on both the explicit and implicit purpose for the design. A key issue to be resolved in future considerations is whether the general purpose for hedonomics should be the promotion of pleasant and enjoyable human–machine interaction, or the goal of promoting well-being and using technology to facilitate well-being. This, in turn, depends in part on whether one's approach to the study of well-being is based on pleasure (hedonism) or self-actualization. While the two goals will in many cases be concordant, there will also be many instances in which they will be contraindicated (i.e., increasing the alienation of the individual and exchanging short-term pleasure for long-term well-being). In such cases, which position prevails will ultimately depend on the intention of those who create and operate the technology. An example may be derived directly from the recent explosion in virtual game-playing. There are a growing number of individuals for whom these virtual interactive games have much greater appeal than "normal reality." Thus, their virtual lives are evolving and interesting; their everyday lives are boring and unexceptional. Are we to create and facilitate their virtual hedonism? That is, is technology to be designed and fostered to support these surrogate lives? There are even some reports, admittedly anecdotal, that perhaps one of these "players" was dying at the keyboard from his/her neglect of the present reality. Regardless of whether this is an urban legend, or whether this is, in fact, a real case of death from "virtual" neglect, it highlights the contrasts between the hedonomic purpose of technology and the goals it supports, whether from the community in, or the user out.

The hierarchy of needs metaphor

The initial theoretical framework through which to promote the opportunity for pleasurable interaction is based on a design priority hierarchy (see Hancock *et al.*, 2005). Maslow's (1970) model of the optimization

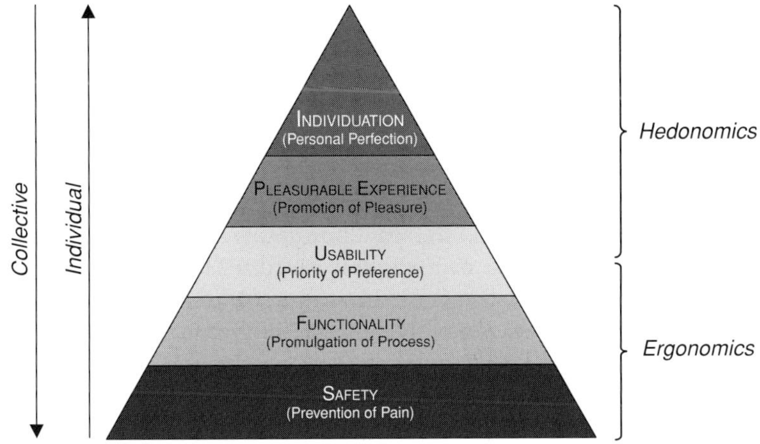

Figure 5.1 The Hedonomic Pyramid (after Hancock, Pepe, and Murphy, 2005).

of human satisfaction through a hierarchy of needs specifies that higher-level needs can be fulfilled only after lower-level needs are satisfied. Hancock and his colleagues took this principle as a design imperative for the ergonomics–hedonomics hierarchy of needs, which is shown in Figure 5.1.

The foundation of this hierarchy is safety. Thus, those who are in peril cannot be expected to reach for, or be concerned with, either pleasure or individuation. Consequently, the primary necessity is to ensure the well-being of the user is secured first in safe operating conditions. The next level requires a functional system that allows the user to accomplish his or her functional goals. Safety and functionality are basic needs on which higher aspirations are based. Once the functionality is achieved, usability facilitates performance by consistently enabling the user to accomplish the task. This experience conditions the user to develop a sense of trust and perceive that the system is consistent, predictable, and safe. In addition, usability improves the system by making it more memorable, efficient, and easy to use, which are all basic elements for pleasurable interaction. When the system achieves usability, it can begin to fulfill users' more psychological and sociological needs, such as their need to achieve, their need to belong, making the interaction a pleasurable experience. To fulfill the users' needs, incorporating an explicit recognition of motivation, users' psychophysiological state, and enjoyment are then necessary in the system design. Thus, the final goal of a hedonic-designed system is to permit each individual user to adapt the system to his or her individual traits, personal views, changing goals, and mood states over time. Some may argue that this pyramid presentation is already outdated in describing the individual person as these two mechanisms for functionality and affect, the intuitive and cognitive systems, act in parallel (see, e.g., Kahneman, 2002).

However, one general question is whether industry and workplaces will adopt hedonic design. From a practical perspective, this pyramid of needs representation reflects the way industry and workplaces could adopt hedonic design. Thus, over the last century safety, ergonomics, and usability were ubiquitously adopted by industries and manufacturers after they were proven to increase productivity and profit. Furthermore, they build upon each other and represent a hierarchy of needs in the workplace, that is, safety is the most fundamental need and usability has been the latest "top level" of the pyramid. Unfortunately, operators' well-being was, and still remains, rarely a matter of concern per se. Human factors have traditionally followed the technology- and performance-driven path by facilitating better designs for systems and workplaces to achieve higher productivity goals. Similarly, hedonomics will not be adopted by industries and workplaces unless it proves to follow this productivity's increasing trend and become the next novel, innovative top hierarchy of the pyramid.

From functional usability to hedonomic fulfillment

To portray hedonomic affordances invites emotional reactions from the user by way of generating a user appraisal that is perceived as something enjoyable, pleasurable, and fun. Traditional cognitive approaches to product usability tend to underestimate the role of user emotions. Not surprisingly, the success of a product in the marketplace may be determined by its aesthetic appeal, the pleasure it creates, and the satisfaction it brings to the user. Such user experience encompasses emotions and expectations in relation to the context of use of the product. Emotions influence how a user interacts with the product. In the interaction, emotions accompany thinking.

A major research problem in the science of hedonomics is to develop methods for measuring emotions. Rarely has the issue of customer emotions been investigated as part of the product development. The literature has documented various methods derived from psychology and marketing to measure emotions (Jordan, 2000; Seva *et al.*, 2007). However, only a few are relevant to product development (Khalid and Helander, 2006). Subjective and objective methods for evaluating user emotion regarding an artifact are currently being refined (see, e.g., Desmet *et al.*, in press). Measurement of emotions may be particularly difficult due to the indefinite relationship between an emotion and its behavioral expression. This is because the mechanics of hedonomics are not fully understood. Further research is needed to develop expressions of emotions that are quantifiable, so that they can be easily verified. Separating emotion from cognitive functions does not seem helpful from a research or design perspective. Instead, an integrated view of emotion and cognition is taking hold, not only in neuroscience, but also in product design. Additionally, a product should be designed to support customer needs, including the customer's persona or personality attributes. This can be done by providing flow – or ease of use – and inducing feelings – or emotions – in interacting with the product. In the future, when we understand the "physics" of emotions and what makes people happy, there will be significant rewards in terms of monetary benefits for developers of pleasurable products, as well as many "feel good" customers. One such example of an area that has evolved tremendously over the past few years is the study of aesthetics and the development of "design for aesthetics" tools. We will use aesthetics here as one example of the many areas where systematic research is necessary in order to generate better hedonic-design guidelines.

Thus, hedonomics is:

- Not – robotic slavery and the end of human work
- Not – a zero level of workload and stress
- But is – challenging, exploratory, featuring discovery, self-determination (autonomy), skill, satisfaction, and human–machine symbiosis.

Three specific areas of research that are directly related to hedonomics and scientific progress are individuation, aesthetics, and etiquette. We have chosen here to briefly present these three current trends.

Individuation

Starting from the premise that emphasizes commonalities across all human beings, we seek to extend such concern to an individual-based design or state of individuation in which the focus is on the unique characteristics of each and every single individual user (see Hancock, 2003; Hancock, Hancock, and Warm, 2009). We believe technology has now progressed to a point that it is capable of dealing with users on this individual, personal level, and this progress fits hand-in-glove with the growing concerns for affect as embraced by hedonomics. Individuation is not just the pursuit of self-centeredness. This is so because optimization of an interface for a specific user alone, which allows him or her to interact intuitively and personally, does not affect only the user; it also affects the user's environment and all aspects of that environment, including other people (see Hancock, 2003). This widespread effect could be either pleasant or detrimental in nature, given that satisfying one individual's hedonomic need has no necessary link to the hedonic state of any other individual. For example, Moray (1993: 35) commented on the benefit of smart houses for "making life easier and safer for inhabitants and reducing energy consumption." The ability of a smart house to detect and adapt to the individual characteristics of its inhabitants would not only allow for individual comfort and optimal operation, but also could conserve energy for the greater good: in this case, a collective positive benefit (see Mynatt *et al.*, 2004). Additionally, human factors/ergonomics specialists have a responsibility to consider potential negative effects of the outflow of poor design and the failure that is produced when we do not take into consideration the strengths and limitations of each human (Vicente, 2004).

Aesthetics and etiquette

It has been argued that modern design has placed too much emphasis on performance issues and not enough on aspects, like aesthetics, that promote pleasure (e.g., Norman, 2003). Studies have shown that aesthetic perceptions influence subsequent judgements of any artifact. Users form immediate aesthetic impressions of web pages, for example, and these impressions are highly stable (Lindgaard, 2007; Tractinsky *et al.*, 2006). Recent research on the visual aesthetics of computer interfaces found that aesthetics are highly correlated with the system's perceived usability both before (Tractinsky, 1997) and after (Tractinsky *et al.*, 2006) the interaction, as well as with user satisfaction (Tractinsky *et al.*, 2006). In a study on web sites, Tractinsky *et al.* found that users have different tastes and reactions to the aesthetics of the web sites. They also found that positive immediate impressions of the web sites were associated with high levels on both classic and expressive aesthetic dimensions. Attractive web pages were associated with the two aesthetic dimensions. However, unattractive pages were associated mainly with very low levels of expressive aesthetics.

Aesthetic standards for web sites may differ from those required for other types of user interfaces. For example, with regard to the aesthetics of in-vehicle navigation displays, Oron-Gilad and her colleagues (cf. Lavie *et al.*, in press) found that only the classical aesthetics dimension was positively correlated with map usability. Thus, more systematic research is necessary in order to fully develop an understanding of the contribution of aesthetics in various human–computer contexts to the fulfillment of hedonic needs.

Etiquette in human–machine interaction is another aspect of design that has been suggested by Miller and his colleagues (2002; Parasuraman and Miller, 2004) as a method of providing more than traditional usability or human-centered design. This evolving branch of research is drawing from culturally familiar patterns of expectations about appropriate behaviors – between humans and machines (e.g., Parasuraman and Miller, 2004; Preece *et al.*, 2002) as machine and automation capabilities become more complex and more sophisticated.

Conclusion

Hedonomics has two main purposes: the promotion of pleasant and enjoyable human–system interaction, and the promotion of well-being and use of technology to facilitate well-being. These two goals may sometimes be contradictory. For example, the question of enjoyment is one that is stable neither in space nor in time. For one to experience epochs of joy or high levels of enjoyment, they must necessarily be contrasted with intervals that are, by definition, less enjoyable. Thus, hedonomic experience is fundamentally cyclic in nature. In the modern 24/7/365 society, we have tended to forget these inherent cycles, instead adopting a continuous temporal ganzfeld (complete field; total) of experience. Worse, we have often taken those very festivals and holidays which have connoted past pleasures and have sought to expand upon these by extending them in time. In so doing, we have inevitably diluted the experience. Let us take, for example, some typical Western "holiday" periods, the most obvious being Christmas, the reflection of the old celebration of the mid-winter solstice. Even within living memory, Christmas was a two to three day celebration with its own unique characteristics – being Christmas Eve and Boxing Day, respectively. However, due to the exploitation of the season by commercial enterprises, Christmas can start (at least in the United States) in early November (if not before). Indeed, as one looks at the annual calendar, one can begin to see that commercial concerns have arranged the dilution of experience, such that there is nearly always a "celebration" at hand. Thus, again in the US, this proceeds from Halloween to the Thanksgiving break to the Christmas interval (almost immediately followed by the now legendary January sales and Valentine's Day, and so on). Thus, in dissolving these peaks of expectation and assumed enjoyment in time, the modern propensity for vast greed has dissolved the fundamental nature of the experience. It is not merely the content of an interval, but its length and the degree of anticipation involved in its arrival, that dictate the emotional response engendered.

Another way to look at the issue of the ganzfeld (completeness) of experience is not to separate the work from the leisure as many people tend to think. The oversimplified categorization of work as something that can not be pleasurable while leisure is all about pleasure is crude and nonproductive. An alternate way to examine the things we do over a day is to distinguish between routine tasks and challenging tasks (regardless of where they are performed). With this perspective, one accepts that not every task done at home can be pleasurable and, vice versa, tasks done at work can be affectively engaging. Once this type of distinction is clearer, job designers can develop better job descriptions which combine the two types of tasks to generate higher fulfillment at work. Furthermore, design criteria for routine tasks may be different than for those affectively engaging ones.

Finally, the role of ergonomics in job design has often been based on a pull (on-demand) approach rather than on a push (leading) one. Hence, ergonomists were hardly ever leading the progress of technology, but rather setting *post hoc* procedures attempting to adapt technology to the task at hand. Now is a good time for ergonomics (and more so hedonomics) to take the lead and become proactive and influential in the development of tools and technology (see, e.g., Oron-Gilad and Hancock, 2004). Industry is not homogeneous in its position toward workers' well-being. The wide range of preferences, goals, and priorities among the various industrial environments is a difficult barrier to scale. For certain types of work domains, there is still a long way to go before the basic hierarchical needs of the worker will be maintained, more so the higher hedonic aspirations. Nevertheless, those hedonic principles should lead our aspirations.

References

Blythe, M. A., Overbeeke, K., Monk, A. E., and Wright, P. C. (2003). *Funology: From usability to enjoyment*. Boston: Kluwer.

Chapanis, A. (1988). Some generalizations about generalization. *Human Factors*, 30(3), 253–267.

Craik, K. J. W. (1948). Theory of the human operator in control systems: II. Man as an element in a control system. *British Journal of Psychology*, 38, 142–148.

Desmet, P. M. A., Karlsson, M. A., and van Erp, J. (in press). *Design and Emotion Moves*. Cambridge: Cambridge Scholars Publishing.

Endsley, M. R. (1995). Toward a theory of situation awareness in dynamic systems. *Human Factors*, 37(1), 32–64.

Fee, E., and Brown, T. M. (2001). Preemptive bio preparedness: Can we learn anything from history? *American Journal of Public Health*, 91, 721–726.

Flach, J. M., and Dominguez, C. O. (1995). Use-centered design – Integrating the user, instrument, and goal. *Ergonomics in Design*, 3(3), 19–24.

Frøkjær, E., Hertzum, M., and Hornbæk, K. (2000). Measuring usability: Are effectiveness, efficiency, and satisfaction really correlated? In *Proceedings of the ACM CHI 2000 Conference on Human Factors in Computing Systems* (pp. 345–352). The Hague, The Netherlands, April 1–6. New York: ACM Press.

Gantt, Henry L. (1913). *Work, Wages, and Profits*. Engineering Magazine Co.

(1919). *Organizing for Work*. New York: Harcourt, Brace, and Howe.

George, J. M., and Jones, G. R. (1997). Experiencing work: Values, attitudes, and moods. *Human Relations*, 50, 393–416.

Gibson, J. J. (1966). *The Senses Considered as Perceptual Systems*. Boston: Houghton Mifflin.

(1979). *The Ecological Approach to Visual Perception*. Boston: Houghton Mifflin.

Gilbreth, L. M. (1956). *Management in the Home*. New York: Dodd/Mead.

Graham, L. (1997). Beyond manipulation: Lillian Gilbreth's industrial psychology and the governmentality of women consumers. *The Sociological Quarterly*, 38(4), 539–565.

Hancock, P. A. (1997). *Essays on the Future of Human–Machine Systems*. Eden Prairie, MN: BANTA Information Services Group.

(2003). Individuation: Not merely human-centered but person-specific design. Paper presented at the 47th Annual Meeting of the Human Factors and Ergonomics Society, Denver, CO, October.

(2009). *Mind, Machine, and Morality*. Farnham: Ashgate.

Hancock, P. A., Flach, J. M., Caird, J., and Vicente, K. (1995). *Local Applications of the Ecological Approach to Human–Machine Systems*. Hillsdale, NJ: Erlbaum.

Hancock, P. A., Hancock, G. M., and Warm, J. S. (2009). Individuation: The N = 1 revolution. *Theoretical Issues in Ergonomic Science*, in press.

Hancock, P. A., Pepe, A. A., and Murphy, L. I. (2005). Hedonomics: The Po"W"er of positive and pleasurable ergonomics. *Ergonomics in Design*, 13(1), 8–14.

Hancock, P. A., and Szalma, J. L. (2008). *Performance Under Stress*. Aldershot, UK: Ashgate.

Helander, M. G. (1997). Forty years of IEA: Some reflections on the evolution of ergonomics. *Ergonomics*, 40(10), 952–961.

Helander, M. G., and Khalid, H. M. (2005). Underlying theories of hedonomics for affective and pleasurable design. In *Proceedings of the Human Factors and Ergonomics Society 49th Annual Meeting* (pp. 1691–1695). Santa Monica, CA: Human Factors and Ergonomics Society.

Illich, I. (1973). *Tools for Conviviality*. New York: Harper & Row.

International Ergonomics Association. (2007). IEA website mission statement. Retrieved July 3, 2009 from www.iea.cc/browse.php?contID= what_is_ergonomics.

Jastrzebowski, W. (1857). An outline of ergonomics or the science of work based upon the truths drawn from the science of nature. Originally published in *Nature and Industry*, No. 29 and No. 30 (1857); reprinted by Central Institute for Labour Protection, Warsaw, 1997.

Jordan, P. W. (2000). *Designing Pleasurable Products*. London: Taylor & Francis.

Kahneman, D. (2002). Maps of bounded rationality: A perspective on intuitive judgment and choice. Nobel Prize Lecture, December 8. Retrieved March 16, 2008, from http://nobelprize.org/nobel_prizes/economics/laureates/2002/kahnemann-lecture.pdf.

Khalid, H. M. (2004). Conceptualizing affective human factors design. *Theoretical Issues in Ergonomics Science*, 5(1), 1–3.

Khalid, H. M., and Helander, M. G. (2006). Customer emotional needs in product design. *Concurrent Engineering*, 14(3), 197–206.

Kira, A. (1966). *The Bathroom*. Center for Housing and Environmental Studies. New York: The Viking Press.

Lavie, T., Oron-Gilad, T., and Meyer, J. (in press). Usability and aesthetics of map displays. *International Journal of Human–Computer Studies*.

Lazarus, R. S. (1991). *Emotion and Adaptation*. Oxford: Oxford University Press.

Lindgaard, G. (2007). Aesthetics, visual appeal, usability, and user satisfaction: What do the user's eyes tell the user's brain? *Australian Journal of Emerging Technologies and Society*, 5, 1–16.

Locke, E. A., and Latham, G. P. (1990). *A Theory of Goal Setting and Task Performance*. Englewood Cliffs, NJ: Prentice-Hall.

Maslow, A. (1970). *Motivation and Personality*, rev. edn. New York: Harper & Row.

Meister, D. (1999). *The History of Human Factors and Ergonomics*. Mahwah, NJ: Lawrence Erlbaum.

Miller, C. (2002). Definitions and dimensions of etiquette. In Christopher Miller (Ed.), *Etiquette for Human–Computer Work: Papers from the 2002 Fall Symposium* (pp. 2–8). Technical Report FS-02-02. Menlo Park, CA: American Association for Artificial Intelligence.

Moray, N. (1993). Technosophy and humane factors: A personal view. *Ergonomics in Design*, 1(4), 33–39.

Murphy, L. L., Stanney, K., and Hancock, P. A. (2003). The effect of affect: The hedonomic evaluation of human–computer interaction. In *Proceedings of the Human Factors and Ergonomics Society 47th Annual Meeting* (pp. 764–767). Santa Monica, CA: Human Factors and Ergonomics Society.

Mynatt, E. D., Melenhorst, A. S., Fisk, A. D., and Rogers, W. A. (2004). Aware technologies for aging in place: Understanding user needs and attitudes. *IEEE Transactions on Pervasive Computing*, 3(2), 36–41.

Nielsen, J. (1993). *Usability Engineering*. Boston: Academic Press.

Norman, D. A. (2003). *Emotional Design: Why we love (or hate) everyday things*. New York: Basic Books.

Oron-Gilad, T., and Hancock, P. A. (2004). Interacting with Nano systems: Is the time right to look for new applications systematically? *Proceedings of the 1st Annual Florida Tech Transfer Conference*, St. Petersburg, FL, May. Retrieved March 16, 2008, from www.mit.ucf.edu/Nanotechnology/OronGilad_Hancock_2004.pdf.

Parasuraman, R., Hancock, P. A., Radwa, R., and Marras, W. (2003). Defending the independence of human factors/ergonomics science. *Human Factors and Ergonomics Society Bulletin*, 46(11), 1, 5.

Parasuraman, R., and Miller, C. (2004). Trust and etiquette in high-criticality automated systems. In "Human–Computer Etiquette," a special section of *Communications of the ACM,* April, C. Miller (Ed.), pp. 51–55.

Powers, W. T. (1973). *Behavior: The control of perception.* Chicago: Aldine.

Preece, J., Rogers, Y., and Sharp, H. (2002). *Interaction Design: Beyond human–computer interaction.* New York: Wiley.

Ramazzinni, B. (1940[1713]). *De Morbis artificum diatriba.* 1700–1713 (W. C. Wright, trans.). Chicago: University of Chicago Press.

Scherer, K. R. (1999). Appraisal theory. In T. Dalgleish and M. Power (Eds.), *Handbook of Cognition and Emotion* (pp. 637–664). Chichester: Wiley.

Seva, R. R., Been-Lirn, H., and Helander, M. G. (2007). The marketing implications of affective product design. *Applied Ergonomics,* 38(1), 723–731.

Smith, K., and Hancock, P. A. (1995). The risk space representation of commercial airspace. Paper presented at the Eighth International Conference on Aviation Psychology, Columbus, OH.

Snyder, C. R., and Lopez, S. J. (Eds.) (2002). *Handbook of Positive Psychology.* Oxford: Oxford University Press.

Taylor, F. W. (1911). *The Principles of Scientific Management.* New York: Harper Bros. In *Internet Modern History Sourcebook,* © Paul Halsall, July 1998. Retrieved March 16, 2008, from www.fordham.edu/halsall/mod/1911taylor.html.

Tractinsky, N. (1997). Aesthetics and apparent usability: Empirically assessing cultural and methodological issues. In *Proceedings of the SIGCHI Conference on Human Factors in Computing Systems.* Atlanta, GA. New York: ACM.

Tractinsky, N., Cokhavi, A., Kirschenbaum, M., and Sharfi, T. (2006). Evaluating the consistency of immediate aesthetic perceptions of web pages. *International Journal of Human–Computer Studies,* 64(11), 1071–1083.

Vicente, K. (2004). *The Human Factor: Revolutionizing the way people live with technology.* New York: Routledge.

Vroom, V. H. (1964). *Work and Motivation.* New York: Wiley.

6 "Good teleworking": under what conditions does teleworking enhance employees' well-being?

Ellen Ernst Kossek, Brenda A. Lautsch, and Susan C. Eaton

Myth: The home is a benign place to work and enables teleworkers to "have it all" – excel at both work and family. Work is portable to home without any downside.

Reality: For many employees, moving the workplace to the home can hurt an employee's home life and create new social dilemmas, especially if employees do not perceive control over the timing, location, and process of work, and believe they are able to create boundaries that allow them to sometimes separate work and family roles.

Given the increasing availability of formal and informal access to teleworking and other flexible work arrangements, in this chapter our goal is to understand the kinds of psychological working conditions under which use of teleworking is more likely to enhance employee well-being. We will show that "good teleworking" or teleworking where use is likely to be related to favorable outcomes for the workers' well-being has several psychological job design characteristics. We draw on survey and interview data from a sample of 316 professional employees in two Fortune 500 firms. Our chapter emphasizes two main features of good teleworking.

First, perceived psychological control over when, where, and how one teleworks matters for well-being. We will show that employees with higher perceived personal control over the location, timing, and process of work experience significantly lower work–family conflict, turnover intentions, and are less likely to want to move to a new career.

Second, how an employee manages psychological and physical boundaries between work and family roles is another critical influence on

Prior to her death Dr. Susan C. Eaton was an important member of our teleworking research group. We would like to acknowledge her significant contribution to the early stages of this project.

We are grateful to the School of Labor and Industrial Relations at Michigan State University for providing graduate assistantships to support data collection and analysis, and especially to Casey Schurkamp for serving as project manager of this study. Correspondence should be addressed to: Ellen Ernst Kossek, kossek@msu.edu. Portions of this original chapter draw on Kossek *et al.* (2006).

whether the use of teleworking enhances well-being. We will show that a boundary-management strategy favoring separation of work and family is significantly related to lower family-to-work conflict. Other types of flexible job design, including access to formal telework arrangements, working in multiple locations, and schedule irregularity, will not significantly relate to lower turnover, career movement preparedness (cf. Kossek *et al.*, 1998), or work–family conflict. Our chapter will argue that mere formal access to teleworking is a necessary but insufficient condition to enhance professional workers' well-being. What matters more is when employees psychologically perceive they have higher personal control over work schedules, process, and location, and when workers believe they are able to choose to create boundaries to separate work and family roles. For example, are workers able to set time boundaries when they can focus on or be available only to family and not to their job colleagues, as well as the reverse? This perceived ability to separate is important as it enables employees to not have role overload through multitasking at times that the employee would rather focus on one role – either their job, the family, or other nonwork roles.

Theoretical background

Early work–family research viewed work and home as independent systems, where a structural and emotional separation of work and family evolved for individuals and organizations (Homer, 2002). Given current shifting labor force composition toward growing numbers of employees managing caregiving and other nonwork demands, more recently scholars have argued that the enforced separation of work and family spheres in traditional workplaces leads to work–family role conflict (Friedman *et al.*, 1998; Kanter, 1977). As potential solutions, researchers and practitioners advocate that employers adopt flexible work arrangements such as flexible schedules and teleworking to help employees integrate and restructure work and family roles to reduce conflicts (Golden, 2001). National surveys show that 84 percent of major employers have adopted flexible schedules and nearly two-thirds (64%) offer telecommuting (Alliance for Work/Life Progress, 2001) with these policies most accessible to professional employees (Bureau of Labor Statistics, 2000). However, questions remain concerning whether the integration of work and family reduces conflicts, whether flexible work arrangements, namely teleworking, provide the intended benefits to the synthesis of work and family roles, and, in particular, whether flexible work schedules and teleworking practices provide professional workers the control they may need to jointly manage often unbounded work hours with nonwork demands.

We define professional workers as individuals with at least a college-level degree who are not paid for overtime and exempt from the Fair Labor Standards Act. They have occupational norms to work as long as it takes to get the job done and usually have some control over starting and stopping times for work and do not punch a clock.

In a classic meta-analysis on flexible work arrangements, Baltes *et al.* (1999) argued that the positive effects of flextime for employees reported in the literature generally did not carry over to professional workers. These findings and other reviews (Avery and Zabel, 2001) suggest some writings in the work–family field have not sufficiently examined whether all forms of flexibility are equally beneficial, nor have they considered whether they benefit professionals – many of whom are typically in jobs that have norms favoring long and unlimited work hours.

Applying Hackman and Oldham's (1980) job characteristics theory, Baltes and his coauthors argued that one important mechanism by which flexibility programs generate positive outcomes is through shifting job designs to enhance worker autonomy or control. Professional workers, they argued, benefit least from flexible work arrangement programs because their jobs incorporate substantial autonomy prior to any shift to an alternative work schedule. However, the Baltes *et al.* (1999) study was not able to test this perspective in the meta-analysis because the study did not include measures of job control. Nor were there studies to examine whether variation in the nature of job flexibility predicted varying work and family outcomes – that is, whether all types of flexibility were equally beneficial for professional employees.

The purpose of this chapter is to integrate and extend previous research on flexible job schedules, design, and teleworking in order to understand how they may relate to work and family outcomes; to distinguish between the effects of availability and use of different types of flexibility, which we argue could be more clearly assessed in the work–family literature on flexible work arrangements; and to investigate relationships to individual boundary management and personal job control. Specifically, we examine how professional workers, who have the greatest access to flextime and flexplace job designs (Kossek, 2006), might benefit most from them. We investigate variation in the features of flexible jobs, including the degree to which jobs are designed to increase personal control over the timing, location, and process of work, and the effect of these characteristics on work and family outcomes.

Managing boundaries between work and family roles is a critical aspect of the control issues we see influencing the effects of flexible work arrangements in general and teleworking specifically. In particular, despite flexible job designs, managers may still attempt to exert "boundary control," which Perlow (1998) has explained includes various efforts to induce professional workers to work longer hours and other actions that shape the boundary between work and family time. Once professional work extends into the home, managers may begin to call workers at home or expect them to answer e-mails in the evenings and on weekends, for example. Introducing work into personal times and spaces also raises the potential of work being increasingly shaped by the employee's management of boundaries in relation to the family role. Recent developments in boundary theory (Ashforth *et al.*, 2000; Clark, 2000) highlight the fact that integrating work and family in time and space, as in flextime and flexplace job designs, means that borders between the two domains are permeable; work may be more interrupted by family influences and vice versa.

For professional workers to really have control over when they shift from work to home or other nonwork responsibilities may not only require a job design that provides autonomy, but also a boundary-management strategy, which Kossek, Noe, and DeMarr (1999) define as the principles one uses to organize and separate role demands and expectations into specific realms of home (e.g., dependent caregiving) and work (i.e., doing one's job). Some professionals with jobs that allow flextime and telecommuting may also desire a segmentation boundary-management strategy; that is, they may seek to establish boundaries between work and home by, for example, setting their own work hours and turning off their cell phone or pager at the end of the day, not checking e-mail in the evenings or weekends, or by working in a home office with a door to shut out family interruptions. More research is needed on the effects of these strategies in the context of flexible job schedules and telecommuting practices, which this study was designed to address.

Variation in the nature of flexibility: implications for work and family outcomes

Work–family researchers have recommended flexibility practices to improve a range of work and family outcomes including work–family conflict (Kossek, 2006), performance (Pierce *et al.*, 1989), intention to turnover, and preparedness to move to a new job (Rau and Hyland, 2002; Scandura and Lankau, 1997). Empirical evidence on relationships has been mixed, but most commonly flexibility practices are found to have some positive effects on both individual and organizational outcomes. (See, e.g., Mokhtarian *et al.* [1998] for a study in which flexible work decreases work–family conflict; Hill *et al.* [1998] for the opposite result; and Baltes *et al.* [1999] for a positive assessment of overall effects across multiple studies.)

However, several reviews have noted weaknesses in the current literature that have hampered progress in understanding the effects of flextime and telecommuting programs and suggest more research is needed before these results can be viewed with confidence as definitive. Empirical studies are often atheoretical (Baltes *et al.*, 1999) and sometimes lack methodological rigor by overrelying on the same source or anecdotal data rather than on statistical analysis or control groups, making it difficult to overcome a positive bias toward the effects of using flexibility in work–family programs (Gottlieb, Kellowy, and Barham, 1988). Research has also underexamined the reality that employees often use different and multiple forms and amounts of flexibility and that access does not necessarily capture use (Bailey and Kurland, 2002).

Accordingly, we focus on a number of types of flexibility that have been identified in prior research, and we adopt a control perspective to understand the effects of flexible work, combining insights from job characteristics theory with theory on control of work–family boundaries. In particular, we argue that it is personal autonomy or control over the timing, location, and process of work that will have the most positive effects for professional workers, rather than other dimensions of job flexibility, and that in fact some types of flexibility may even have negative effects on work–family conflict.

Traditional job characteristics theory predicts that autonomy or control, along with four other core job characteristics, will lead to enhanced motivation, satisfaction, and performance (see Fried and Ferris, 1987, for a meta-analysis supporting the basic propositions of the theory). Further, Baltes *et al.* (1999) have argued that it is only flexible job designs which provide workers with increased control over how the work is done that will lead to better individual and organizational outcomes. We agree with this argument, but note that it is based on a traditional view of control, which focuses on personal job autonomy in how the work is done at the workplace (Hackman and Oldham, 1980).

Recent changes in technology and the nature of professional work are making other forms or aspects of control and autonomy important. Apgar (1998) has noted that professional work has become increasingly portable because of growing use of cell phones, e-mail, and laptops. We theorize that individual control over where and when one works are additional key aspects of job autonomy that should be assessed to update measures of autonomy within work environments. Such an approach captures the growing trend where work can increasingly be done away from the main workplace at different times of the day. In sum, we predict that designing flexible jobs to give professional workers control over not only the process, but also the timing and location of work, will enhance effectiveness. We define this as *personal job flexibility control*, which is control over where, when, and how one works. We surmised professionals with personal job flexibility control may experience lower work–family conflict since they will be able to restructure work and family demands as needed. Their performance may be enhanced since they will experience fewer work or family interruptions due to the ability to rearrange roles as needed. They also will be more likely to want to stay in their job and in their career, as the ability to control work hours is generally highly valued by skilled professional workers. They will have lower career movement preparedness (Kossek *et al.*, 1998), which is defined as self-management behaviors such as feedback seeking and networking to have readiness to change careers, since they will find their current career more satisfying due in part to the ability to have higher personal job flexibility control.

Hypothesis 1: *Personal job flexibility control* will be positively related to performance and negatively related to work-to-family conflict, family-to-work conflict, turnover intentions, and career movement preparedness.

We next consider four other aspects of flexibility that have previously been considered in the literature: formal access to flexible work programs, amount of use of flexibility practices, schedule irregularity, and working in multiple locations. Work–family researchers have argued, based on social exchange theory (Blau, 1964), that formal access to supportive human resource policies like telecommuting will lead to more positive outcomes as employees who value the policies reciprocate with improved workplace attitudes and behaviors (Allen, 2001; Grover and Crooker, 1995; Roehling *et al.*, 2001). However, recent research shows the importance of not confounding access and use. Allen (2001) found that greater use of work–family policies providing flexibility such as telecommuting, flextime, and compressed weeks is related to less work–family conflict, while mere formal availability is not. Eaton (2003) also found that usability of work–family policies – that employees feel free to use actual flexibility – is what is critical for positive outcomes, rather than simply availability. As a result, we hypothesize limited effects on work–family effectiveness for this dimension of flexibility (*formal access*).

Hypothesis 2: *Formal access* to telework will not be significantly related to work–family effectiveness, including work-to-family conflict, family-to-work conflict, turnover intentions, career movement preparedness, or performance.

Given the generally positive results across many studies for flexibility programs, individual use of flexible job designs is expected to generate positive effects across a range of work and family outcomes. However, in examining use of flextime and flexplace practices, much of the literature has framed use of flexibility as a dichotomous variable. For example, if one used a flextime program, writings in the work–family literature seemed to assume that the individual had a flexible job and if they did not, then they had an inflexible job. We argue that research should also consider the *extent* to which jobs are flexible, and whether results will be uniformly positive across all varying degrees of use and for all outcomes.

Baltes *et al.* (1999) find that highly flexible flextime programs (i.e., those with fewer core hours during which employees were required to be in the office, and greater numbers of flexible hours during which employees could determine their start and end times) were less effective in terms of work-related outcomes. Too much flexibility, they argue, may create coordination problems for the worker and colleagues and be simply too complex to manage. Supervisors may rate the performance of workers who are heavy users of flexibility lower than others, since they may be seen as increasing managers' need to coordinate work (Kossek, Barber, and Winters, 1999). We predict, then, that while use of flextime and telecommuting practices is likely to have beneficial effects on individual psychological outcomes, including work–family conflict (both directions), turnover intentions, and career movement preparedness, those who use higher *volumes of flexibility* are likely to receive lower performance ratings by their supervisors.

Hypothesis 3: *Volume of flexibility* will be negatively related to turnover intentions, career movement preparedness, work-to-family conflict, family-to-work conflict, but also negatively related to performance.

We also conceptualized two variables capturing other types of flexibility that could be significantly related to work and family outcomes: *schedule irregularity* (i.e., frequent changes in daily working hours) and *place mobility* (e.g., working at multiple locations, such as at home, the office, a client's office, and on the road). Given the higher likely number of process losses and transaction costs resulting from shifting between varying work schedules and/or multiple places (Ashforth *et al.*, 2000), we expected these aspects of flexibility might actually worsen individual work–family attitudes and performance.

Hypothesis 4: *Schedule irregularity and place mobility* will be negatively related to performance, and positively related to work–family conflict (both directions), turnover intentions, and career movement preparedness.

Flexible work and controlling the work–family boundary

We argue that preferences for work and family boundaries are socially constructed, and there is some social choice in how individuals define boundaries, as do Ashforth and colleagues (2000). Kossek, Noe, and DeMarr (1999) hold that a boundary-management strategy is part of one's preferred approach to work–life role synthesis. Individuals have a preferred, even if implicit, approach for meshing work and family roles that reflects their values and the realities of their lives for organizing and separating role demands and expectations in the specific realms of home and work. This view is consistent with what Zedeck (1992) argued is at the heart of the issue of work–family balance: the way individuals shape the scope and parameters of work and family activities, create personal meaning, and manage the relationships between families and employees in organizations.

In order to organize their varying work and family roles, Nippert-Eng (1996) suggested that individuals construct mental and sometimes physical fences as a means of ordering their social, work, and family environments. Through ethnographic interviews, she found that some individuals are mainly integrators. These people like to blend work and family roles, switching between baking cookies with the kids and downloading e-mail. Other individuals are mainly separators – they prefer to keep work and nonwork separate, rarely working from home or on the weekends, for example.

Limited empirical research has been conducted on the implications of different boundary-management strategies, in part because the concept is relatively new in the literature. We consider the linkages of boundary-management strategy to types of flexibility used, and work and family outcomes.

Implications of boundary-management strategy

Noting that it is difficult today for growing numbers of employees to perform their jobs without interaction with the caregiving role and vice versa, many work–family theorists argue that greater integration between work and family roles is a way to balance work and family life and even to catalyze positive effects in the other (Friedman *et al.*, 1998; Rapoport *et al.*, 2002). Yet recent theory runs counter to the prevailing belief that integration is generally a "good thing" for individuals. Boundary theory suggests that the increased process losses, role transitions, and transaction costs often associated with role switching from family to work or work to family may not necessarily lead to less interrole conflict

(Ashforth *et al.*, 2000). For example, workers may find their ability to attend to both work and home demands undermined by more frequent interruptions if no limits are placed on times when colleagues could phone home or family members could walk into the home office. Therefore, we predicted higher family-to-work and work-to-family conflict for strategies favoring integration.

Hypothesis 5: *A boundary-management strategy* favoring integration will lead to less favorable work and family outcomes (particularly higher work-to-family and family-to-work conflict).

Method

Participants and procedure

This study relied on a sample of 316 professionals. These employees worked in professional jobs in information technology and systems engineering, communications, finance, marketing, and human resources at two large information and financial services organizations geographically distributed across the US. The firms were similar in work environments and had similar professional job requirements (e.g., writing, e-mail, and use of Internet, programming, phone sales, and project management) where many job tasks could be done as easily virtually as in the formal company office. Over the past few years, both of these firms had growing numbers of professionals who had access to telework during the workday or evenings and weekends.

The sample was well educated: 80 percent of these employees held at least a Bachelor's degree. The sample was 57 percent female and 90 percent Caucasian. About half (48%) had dependent children and 4 percent had at least one elder dependent for whom they regularly cared. Nearly three-fourths (72%) had formal permission to telework. Since some professional workers telework informally, the actual number of workers who report teleworking is even higher: 86 percent and 80 percent of our respondents at the two companies we studied reported doing at least some work away from their main office. Approximately 30 percent of the sample were 35 years of age or younger, 48 percent were between 36 and 45, and 22 percent were 46 years of age or older.

We contacted 626 professional workers at the two companies we studied. Prior to data collection, all individuals signed a voluntary written consent to participate after reviewing the purpose of the study (i.e., to examine the work and family effects of teleworking and job flexibility) and a statement ensuring the confidentiality of all individual results.

Professional workers who consented to participate were sent a survey (either written or e-mailed, as they preferred) covering demographics and job and family background. They subsequently participated in a taped telephone interview that was about forty-five minutes in length. Three months later, interview data on performance ratings were collected from a subsample of ninety participants' supervisors. The response rates for both the employee and the supervisor data collections were 50 percent and 52 percent respectively, with similar response rates at both firms.

Measures *Flexibility type.* Five measures were developed to assess flexibility type: informal personal job flexibility control, formal access to telework, flexibility volume, place mobility, and schedule irregularity. The *informal personal job flexibility control* measure assessed personal freedom to control where, when, and how one did one's job with six items. When we looked at traditional measures of autonomy in job design, we did not see measures of control over the location and timing of work being fully captured as basic elements of job design in widely used measures such as the Job Diagnostic Survey (JDS) (Hackman and Oldham, 1980), which only measures autonomy in how the work is done. Our measure includes two items from the original JDS that capture traditional job autonomy over how the work is done. They were: (1) "How much autonomy is on your job? To what extent does your job permit you to decide on your own about how to go about doing the work?"; and (2) "The job gives me considerable opportunity for independence and freedom in how I do the work." We then constructed four items to capture flexibility control over work location and scheduling, since personal control over when and where one works (i.e., time and place autonomy) are new facets of the personal autonomy construct we assessed. These were "To what extent does your job permit you to decide on your own about WHERE the work is done? To what extent does your job permit you to decide about WHEN the work is done? I have the freedom to work wherever is best for me – either at home or at work; I do not have control over when I work (reverse)." The items were scored on a 1 to 5 Likert-type response scale, with higher numbers indicating more personal job flexibility control. Coefficient alpha reliability for this scale was 74.

Formal access to telework was coded a dummy variable. It was based on employee records provided by the human resource department regarding who had official permission to conduct any part of their work remotely.

Flexible work volume, defined as work done at a distance from the main workplace, was measured with the item, "What percent of your job do you currently do away from your main office or customer?"

Place mobility measured the number of different places that individuals worked during a week. Participants indicated during a typical week whether they worked at home, on site, and/or at other places. Working in one place was coded as 1 (low mobility), two places was coded as 2 (medium mobility), and three or more places was coded as 3 (high mobility).

Schedule irregularity assessed whether an individual tended to have the same predictable daily schedule throughout the week or had a schedule that changed from day to day. To develop this measure, employees were asked to indicate their typical daily work schedule for an average week, using the most recent average week. Respondents with two or more different schedules, defined as differing by at least one hour in starting time, stopping time, or nonwork gaps, were coded as "1" for having significant schedule irregularity, and respondents with the same schedule all week were coded as "0" for having little or no schedule irregularity.

Boundary-management strategy. This measure was derived from Kossek, Noe, and DeMarr's (1999) definition of the construct of boundary-management strategy. It ranged from a strategy favoring high separation, where one strives to keep work and personal roles very separate, to a strategy favoring high integration, where one strives to let work and family roles blur. During the taped interview, individuals were first read the statement, "With the increasing demands of work and home, employees may work in different ways to handle these demands." They were then asked, "All in all, do you currently see yourself as someone who tries to keep work and personal roles separated most of the time, or someone who tries to keep them integrated?" Employee responses were dummy coded (0=separate; 1=integrate).

Individual difference measures. Three variables were included to control for job and individual differences that affect work and family outcomes. If an employee had at least one dependent child, partner, or elder for whom they provided *regular dependent care*, this item was coded as a 1. Otherwise the item was coded as a 0.

Gender was dummy coded: male (coded 0) or female (coded 1).

Total work hours measures the total number of hours the employee typically works in a week.

Dependent variables *Turnover intentions.* This construct was measured with two items developed by Boroff and Lewin (1997). They were: "I am seriously considering quitting this firm for an alternate employer," and "During the next year, I will probably look for a new job outside the firm." These items utilized a five-point Likert-type response scale with higher responses indicating more agreement. The coefficient alpha reliability for this scale was 86.

Career movement preparedness. This construct was measured with four five-point Likert-type items from the scale developed by Kossek *et al.* (1998). Sample items include: "To what extent do you/have you: explored job opportunities on the Internet." The coefficient alpha reliability for this scale was 71.

Work–family conflict. We used a four-item work-to-family conflict scale and a three-item family-to-work conflict scale adapted from Gutek, Searle, and Klepa (1991). Sample items include: "My work takes up time that I'd like to spend with my family and friends," and "My supervisors and peers dislike how often I am preoccupied with my personal life while at work (reversed)." Coefficient alpha reliabilities for the two subscales were 73 and 71, respectively.

Supervisor performance rating. We conducted phone interviews with supervisors for a subsample of our employee sample, and asked them to respond to eight items developed by Fedor and Rowland (1989) stating, "Please rate employee X's overall performance on the following characteristics." The list of characteristics included "Overall performance quality," "Avoiding mistakes," and "Performing up to the supervisor's standards." The higher the score on the five-point Likert-type scale, the better the performance. Coefficient alpha reliability was 91.

Results

Table 6.1 presents the means, standard deviations, and intercorrelations for all variables in this study.

We used ordinary least squares regressions to test our hypotheses. Table 6.2 shows the results for regressions with career movement preparedness, work–family conflict (both directions), and turnover as dependent variables, and Table 6.3 shows the results for the regression with performance as an outcome.

The regression results show that flexibility type significantly predicted work–family effectiveness. We predicted in hypothesis 1 that greater personal job flexibility control would improve professionals' well-being. We found significant support for this idea. While our results show that personal job flexibility control does not affect performance, it does significantly reduce work–family conflict (in both directions) as well as career movement preparedness and turnover intentions. We argued (in hypothesis 2) that mere access to telecommuting would not have these beneficial effects and our results support this: Formal access had no significant relationship to any of our dependent variables. Turning to hypothesis 3,

Table 6.1 *Means, standard deviations, and intercorrelations for all the variables in the study*

		N	Mean	SD	1	2	3	4	5	6	7	8	9	10	11	12	13	14
1	Work-to-family conflict	316	2.81	0.75	1.00													
2	Family-to-work conflict	316	1.78	0.47	0.09	1.00												
3	Supervisor performance rating	84	3.91	0.62	0.02	–0.08	1.00											
4	Career movement preparedness	316	2.36	0.83	0.14*	0.05	–0.22*	1.00										
5	Turnover intentions	316	1.95	0.95	0.18**	0.12*	0.16	0.47**	1.00									
6	Personal job flexibility control	316	3.84	0.77	–0.11*	–0.17**	–0.01	–0.21**	–0.29**	1.00								
7	Formal access to telework	316	0.72	0.45	0.01	0.01	0.29**	–0.21**	–0.16**	0.31**	1.00							
8	Volume of portable work	316	43.24	39.48	–0.11	–0.05	0.22*	–0.26**	–0.17**	0.36**	0.54**	1.00						
9	Schedule irregularity	299	0.66	0.47	0.14*	–0.07	0.05	–0.01	–0.01	0.10	0.12*	0.10	1.00					
10	Place mobility	305	1.65	0.68	0.13*	–0.06	–0.08	0.06	0.03	0.06	0.00	–0.22**	0.22**	1.00				
11	Boundary-management strategy	314	0.58	0.91	0.02	0.13*	0.22*	–0.01	–0.04	0.00	0.10	–0.02	0.05	0.11	1.00			
12	Dependents	310	0.49	0.50	0.03	0.17**	–0.01	–0.08	0.04	0.09	0.12*	0.09	–0.01	0.11	0.07	1.00		
13	Gender	316	0.57	0.50	0.03	0.02	0.29**	–0.06	–0.10	–0.05	0.02	–0.01	–0.10	–0.21**	0.11**	–0.02	1.00	
14	Total work hours	316	45.11	8.25	0.30**	–0.11*	–0.02	–0.07	0.02	0.06	0.07	–0.04	–0.45**	–0.29**	0.03	0.14**	–0.21**	1.00

*$p < .05$, **$p < .01$

Table 6.2 *Results of regression for work and family attitudes*[a]

	Work-to-family conflict	Family-to-work conflict	Career movement preparedness	Turnover intentions
	Beta	Beta	Beta	Beta
Controls				
Gender	.10	–.04	–.05	–.11
Dependents	.09	.17**	–.04	.08
Total work hours	.31**	–.07	.06	.02
Flexibility type				
Personal job flexibility control	–.13*	–.18**	–.14*	–.27**
Formal access to telework	.06	.07	–.08	–.06
Volume of flexible work	–.09	–.05	–.16*	–.05
Schedule irregularity	.02	–.01	–.01	.01
Place mobility	.04	–.08	.02	–.00
Boundary-management strategy	–.02	.14*	–.01	–.03
Total Δr^2	.13**	.09**	.10**	.11**
Note: After pairwise deletion n = 292				

*$p < .05$, **$p < .01$
[a]standardized beta coefficients reported

we predicted that individuals who spent a greater portion of their time telecommuting would benefit in terms of lower work–family conflict and reduced desire to leave their organizations, but that they might pay a penalty in terms of performance as working high volumes at a distance can create coordination problems with colleagues and supervisors. However, our results only partially support these ideas. We found that a higher volume of hours spent telecommuting reduced career movement preparedness, but had no other significant effects on the outcomes we examined. In hypothesis 4 we examined two further aspects of flexible work – the irregularity of individuals' schedules and the extent to which individuals worked in multiple locations. We anticipated that flexible jobs with these characteristics would have negative effects, but, contrary to our expectations, found no significant relationships to work–family well-being.

Table 6.3 *Results of regression for performance*[a]

	Performance rating
	Beta
Controls	
Gender	.27*
Dependents	–.04
Total work hours	.01
Flexibility type	
Personal job flexibility control	–.13
Formal access to telework	.23
Volume of flexible work	.14
Schedule irregularity	.04
Place mobility	–.02
Boundary-management strategy	.17
Total Δr^2	.22*
Note: After pairwise deletion n = 80	

*p < .05, **p < .01

[a]standardized beta coefficients are reported

Hypothesis 5, which predicted that boundary-management strategies focused on integration would lead to worse outcomes for individuals, also received partial support. Boundary-management strategy was a significant predictor of family-to-work conflict. Employees with a boundary-management strategy favoring integration experienced significantly higher family-to-work conflict.

Finally, results for our control variables show that long work hours are positively associated with work-to-family conflict, that responsibility for dependent care increases family-to-work conflict, and that being female and performance ratings are positively associated in our sample. Regarding the latter relationship, given the fact that nearly three-fourths of the sample did some teleworking and often individuals with family caregiving demands have greater interest in these arrangements, it may be that the better-performing women were more interested in and able to get supervisory approval to use flexibility practices.

Discussion: challenging the myth that teleworking necessarily enhances work–family balance

Our study's results challenge the myth that telework necessarily enables work–family balance. There is a great deal of hype oversimplifying the wonders of the virtual workplace. Our results show that higher individual control over the location, timing, and process of work has beneficial effects for work–family conflict, turnover intentions, and career movement preparedness; however, other types of flexible job designs, including access to formal teleworking arrangements, working in multiple locations, and schedule irregularity, may have less positive effects on work and family outcomes. The management literature has overlooked many of these differences in flexible job designs and oversimplified the benefits of making jobs flexible and virtual. A main issue our study raises is that there may be costs associated with flexibility and with moving work into the home, which used to be a place of personal respite and peace. What other problems and conflicts did the workplace of the twenty-first century create and what myths of flexibility need to be debunked or at least examined? For example, for some jobs, flexibility means an employee is available to work 24 hours 7 days a week! In the long run, is this really good for employers, individuals, and their families?

Our results suggest that, for many employees, moving the workplace to the home can hurt home life and create new social dilemmas, especially if workload is not reduced, space and technological infrastructure inadequate, and the family has little understanding of these roles. We also collected some qualitative data that triangulates nicely with our quantitative results noted above. One key subtheme we found was that teleworking changes family and friends' expectations as indicated by some responses below by different teleworkers.

Teleworking changes family and friends' expectations

Here are some quotes from some of the employees we interviewed on problems they were having with managing new social relationships with families and friends while teleworking. These qualitative comments help us delve deeper to better understand our quantitative findings of why a separation strategy of having times when a teleworker was not available to family or friends or work was significantly related to lower work–family conflict for teleworkers.

"They think you're available, and people tend to think you're not really working."

"They expect that I'm available at home. When I'm not, it annoys them."

"My wife is always looking for me to do other things besides work."

"My husband goes out of his way to stay with the schedule. The children have a harder time understanding, though."

"My children call from school more often than when I am at work. My family expects that I can do things like running errands at lunch."

"They think I won't get as much done, my co-workers. My children think I'm available when I'm at home."

"My friends will want me to do lunch with them. My kids are more demanding when I'm at home."

"My kids and even my wife don't understand that my focus is on work."

"They think I'm not working. It's very annoying. That happened more in the beginning."

"Yes, it's hard for them to understand that I'm working for my office and not available."

A review by Kellogg (2001) noted that negotiating boundaries and understanding issues of time and space is an increasingly important issue in the modern organizational and family environment. As we have discussed above in the literature review, Kossek, Noe, and DeMarr (1999) refer to this issue as one of work–family role synthesis: the strategies an individual uses to manage the enactment of work and family roles. It involves decision-making choices governing boundary management and embracement of multiple roles. Some employees are integrators, preferring to mix boundary roles throughout the day; while others are segmenters, preferring to separate roles (Kossek, Noe, and DeMarr, 1999; Nippert-Eng, 1996).

Despite these theoretical papers, very little empirical investigation has been done on how teleworkers actually enact boundary, time, and space negotiation. Future research should build on this study and examine how boundary-management strategies relate to the activities of family members, how they affect an employee's work practices, and how they shape personal and organizational outcomes (Kellogg, 2001).

Sadly, unless employees are able to control when they mix work and family roles in a way that is congruent with their personal preferences for integration or separation, our study suggests that the benefits expected from telework, like having more time with one's children and an easier time meeting family demands, may actually be elusive. For example, children and elders may become confused as they "see mom and dad as physically available," but become annoyed when they are not mentally available. Neighbors may call on the phone and become perplexed when the employee is unavailable. Issues over space and who gets to use the home computer are also raised. Barking dogs and crying babies and other distractions may create problems when clients call a home number. We hope future research will continue to build on this current study and follow up on these issues raised.

Teleworking may lead to overwork

Telecommuting also will not lead to lower work–family conflict or to greater well-being if it increases the risk of individuals overworking. If professionals are unable to set boundaries or limits around their work, telecommuting can lead to colleagues, clients, or supervisors contacting the worker 24–7 and to expectations of constant availability. The "flexibility" in this kind of work arrangement may only or largely benefit the organization and client, and may mean frequent and unpredictable interruptions of family time for the individual. A further problem is that individuals may have trouble "turning work off." Work is a very powerful force as employees must subjugate their personal needs to those of the employer and their financial livelihood. As work is increasingly coming into the home, it may be more and more difficult for individuals to feel that there are times that they can adopt a separation strategy. The comments of telecommuters quoted below illustrate these dynamics.

"I get weekend calls and evening calls. When I'm sick, they [at work] still expect me to get work done since I don't have to come into the office."

"My flexibility includes carrying a pager and understanding interruptions."

For some professionals, this kind of constant availability to work may result in longer work hours. One issue that managers, employees, and organizations must ponder in today's world of growing telework or portable work is "How much work is 'too much'?" At what point should we label an individual's work hours as "overwork" and see it as a problem causing work contagion or "bad teleworking"? The perspectives and interests of employees and employers are likely to differ on this matter. One view is that an employee works "too much" when the employee is unhappy and is showing physical and psychological signs of stress. These might include difficulty separating from work mentally, rarely taking breaks, not getting any exercise at all, and being physically able to unwind.

Another perspective is that an employee works "too much" when the total hours worked by teleworking employees are at least several hours a week more than the total hours worked by others with the same job who are working in the office. We believe that work hours may be labeled overwork when both these factors are combined – when the portable worker is clearly working more hours than their nonportable counterparts and they are unhappy about it.

Why do we have the bar of more than the number of hours usually worked by office workers? One critical issue is who owns the "extra" work hours saved from portable work, such as those saved by not having to commute, not having to shave or put on nice clothes or makeup, and not having to spend much time schmoozing at the coffee stand or waiting to make copies at the copy machine. Also, there is a big difference between working one or two extra hours a week and working ten, fifteen, or more hours beyond what is "normal." A social dilemma that teleworking raises for organizations and employees is the growing lack of clarity in the psychological employment contract regarding mutual expectations of "normal working hours."

A worker could be working two extra hours a week but still have an extra four hours with family beyond those they would have without portable work. This could occur, for example, if six hours were saved by not spending time commuting and chit-chatting at the office, and the worker devoted two of these hours to work and four of these hours to the family. If the worker is happy with their work hours and the company is getting a benefit from portable work, maybe then it is okay if portable workers work a little longer than office workers. The employee is willing to work a little longer and is more focused than when in the office in return for the reduction of stress related to no commute, control over work schedules, and more family and personal time. So to be a win-win option, both the organization and the worker have to be somewhat satisfied with the level of work output – the complication here is to not confuse effort and time (hours per day) with output, especially in jobs where the output is hard to measure, so we evaluate the success of the work *process*. By evaluating the work process, we mean determining whether the way of teleworking is working well for the individual and the organization, and the employee feels they have some control over when and where and how they work – our personal control measure. If the company is not getting its fair share or if the person feels they have little personal flexibility job control, is showing signs of stress or burnout, or has a high work–family conflict level, then the teleworking arrangement is not working. On the other hand, if the company is satisfied and the worker and the family overall feel they are psychologically benefiting from the teleworking arrangement, then this would be a win-win option. It is important for managers and employees (and even the worker with their families) to set aside times at least once or twice a year to discuss whether the telework arrangement is benefiting the worker and the company in a mutual way – that is, "good teleworking" that is win-win for the employee and the employer.

Theoretical and future research implications

The results of our survey and qualitative interviews have important implications for theory and for future research on flexible work and telecommuting. Although leading work–family theorists have suggested that the field needs to shift from a focus on policy alone to the practice and processes of change (Rapoport *et al.*, 2002), few studies have done so to date. We show that those with formal access did not have significantly lower work–family conflict (in either direction), contrary to much of the existing work–family literature. Future research should not confound availability and use or assume that the mere availability of flexibility (such as telework access) is enough to reduce work–family conflict.

Another important finding for future research was that all forms of flexibility are not necessarily good for individual well-being and job performance. We offer new theory on how to conceptualize flexibility, that it includes the availability of formal policies, the type of flexibility used on the job (how much work is done away from the main place of work, how irregular the schedule and mobility, or how many places in which it is done), and the amount of personal control over flexibility use. These measures and concepts should enable future researchers and organizations to better understand the varied nature of flexibility for employees and firms.

Personal job flexibility control was the most important aspect of flexibility for positive employee experiences of lower levels of work–family conflict, lower intention to turnover, and lower career movement preparedness. Thus, we build on classic job design theory (Hackman and Oldham, 1980), which predicts that autonomy or control over the work process will lead to improved employee attitudes and performance, and show that control over the timing and location of work is also important for these outcomes and for other measures of work–family effectiveness. Our findings suggest that formal human resource policies offering flexibility such as teleworking may not help employees manage work and family conflicts, unless employee users also experience an ability to have some job schedule control.

We have moved the workplace into the home for at least part of the work week or evenings and weekends, for many professionals, without enabling workers and managers and families to fully develop new social, cultural, and structural systems to delineate roles and effective coping strategies, supports, and expectations. The work–family literature may have overstated the upsides of flexibility access and "integration" boundary-management strategies, and this research has shown they are not a panacea. Access is a necessary but insufficient condition for higher personal effectiveness in areas such as reducing work and family conflict.

Another important finding of this chapter is that higher family-to-work conflict is associated with an integration-oriented boundary-management strategy. These results are consistent with theory and evidence developed by Ashforth and colleagues (2000) that, contrary to the popular press, an integration of work and family boundaries does not necessarily correspond with less family-to-work conflict. This finding may be due to increased role transitions and process losses from having to switch back and forth and refocus between work and family roles. An integration strategy may also allow for greater permeability between roles. When something good or bad is happening in one domain, it may be more difficult to buffer good or bad things entering the other life space.

Study limitations and practical implications

Despite its strengths, there are several limitations to this study. Although we report results on teleworkers and nonteleworkers, and have performance data collected from supervisors separately from the employee data, a study limitation is that it uses cross-sectional self-report data for some measures. Cross-sectional research, of course, cannot demonstrate direction or causality of effects, so integration boundary-management strategies may be a result rather than a cause of higher family-to-work conflict. Longitudinal research, measuring both family-to-work conflict and boundary-management variables at different points of time, would help to clarify this relationship.

We should also note that we only measured work–family conflict in this study. Well-being is more than just the lessening of negative effects – it can also include positive aspects of well-being, for example balance and enrichment. Future research should also include studies of work–family enrichment and positive spillover from teleworking, which was beyond the scope of the current study.

Scholars could also extend our work by considering other populations of workers, and other influences on boundary-management strategies. Our sample is solely professional with similar kinds of work – future researchers would surely want to broaden the lens to look at more kinds of employees in a wider variety of jobs at all levels of organizations. A final limitation of our study that further research should address is that this study does not explore fully the interplay in how boundaries are enacted not only on an individual level but also as a culturally driven phenomenon. For example, Poster and Prasad (2005) found differences in how professionals in the US and in India had very different cultural norms about boundary management and that workaholism can be as much a function of societal and organizational norms as of individual proclivities.

Future research might also use time diaries and beepers or shadowing in order to more finely measure different kinds of boundary-management strategies and flexible job designs. These methods are very expensive, but may be well worth the investment. Despite these potential areas for improvement, this study adds to our knowledge by examining the mixed effects and multifaceted aspects of flexibility. A clear practical implication of the study is that work–family boundary integration may arise naturally with flexible working arrangements as volume and irregularity of flexible work increases, unless individuals strive to counter this with strategies to segment work and nonwork roles, and organizations allow them to do this. An example of a personal strategy is having a separate door to a home office and hiring a full-time babysitter while working. We have downloaded the office onto some employees' homes, and they and their families may not yet have learned effective strategies to manage these new work arrangements.

Additional studies should build on our research on the construct of boundary-management strategy to further examine how people may shift rhythms over daily, weekly, and lifespan changes and how they are associated with different types of flexible job designs. As noted, the work–family literature places boundary management on a continuum from segmentation to integration, and there may be more complexity to this issue to investigate in future work. For example, if an employee is working at home with the door closed while their child is watching television; some could say they are physically integrating roles; they are working at home and are physically there, but are mentally segmenting as they are not interacting with their family. People cannot move work into the home without changing their social relationships. Future research should develop additional measures of the various aspects of boundaries that are being integrated/separated (physical, mental, behavioral, temporal), the implications of integrating on some parts of the boundary, but not others, and the waxing and waning of the process of boundary management over a workday, work week, and the life course. More research is needed on coping strategies individuals can adopt to help set boundaries that fit with their preferences. Training in negotiation skills might be helpful so that individuals feel empowered to speak up and negotiate flexibility enactment approaches that help not only their work effectiveness but also their personal and family effectiveness. Supervisors also may need additional training on how to better manage and provide more effective support to employees in these transformational work arrangements.

Conclusion

We have written this chapter with the goal of contributing to future studies on the effects of enhanced control and autonomy over one's work characteristics upon personal and organizational outcomes. Unlike many earlier studies, the chapter outlines a theory-based model which is then empirically tested. We argued that, in addition to traditional work design theory and elements, increased autonomy/control over timing, location, and work processing, which is afforded by technological developments, will have main effects on control, and do have positive effects on avoiding or lowering of work–home conflicts (two directional) for professionals.

In this chapter we make a distinction between access and use of flexible arrangements, which has sometimes been confounded in the literature. Our results show that more access to flexibility is not always better (see Kossek and Lautsch's [2008] discussion of flexstyles in their new book, *CEO of Me*). Too high levels of flexibility may yield negative rather than positive outcomes as such flexibility may create new demands that are difficult for many individuals (and families) and organizations to handle without new social learning. A home and work integration approach may be problematic as the two realms become too mixed to be effectively controllable. Having formal access to flexibility is not the same as having control. Even when working at/from home, we are not always alone. There are the expectations, preferences, and time use by others with whom we have to communicate even at/from home. Colleagues, bosses, and clients from the work realm and new expectations and preferences of family members may enlarge and extend work conflicts, home conflicts, and between-work-and-home conflicts. Teleworking as a panacea may, therefore, be an illusion of detachment and extra freedom, not necessarily a reality.

Rapoport and her colleagues (2002) call for integrated action research by scholars and organizations. We concur and hope this chapter will prompt future scholarly work to tease out more about how organizations adopt, distribute, and enact flexibility, the individual and job conditions that lead to the effectiveness of flexibility practices, and develop greater understanding of how flexible work arrangements can provide greater benefit to individuals and organizations.

References

Allen, T. D. (2001). Family-supportive work environments: The role of organizational perceptions. *Journal of Vocational Behavior*, 58, 414–435.

Alliance for Work/Life Progress (2001). Telework trendlines. Retrieved July 3, 2009, from www.worldatwork.org/waw/adimLink?id=21945.

Apgar, M. (1998). The alternative workplace: Changing where and how people work. *Harvard Business Review*, May–June, 121–136.

Ashforth, B. E., Kreiner, G. E., and Fugate, M. (2000). All in a day's work: Boundaries and micro role transitions. *Academy of Management Review*, 25, 472–491.

Avery, C., and Zabel, D. (2001). *The Flexible Workplace: A sourcebook of information and research*. Westport, CT: Quorum.

Bailey, D. E., and Kurland, N. B. (2002). Review of telework research, findings: New directions, and lessons for the study of modern work. *Journal of Organizational Behavior*, 23, 383–400.

Baltes, B. B., Briggs, T. E., Huff, J. W., Wright, J. A., and Neuman, G. A. (1999). Flexible and compressed workweek schedules: A meta-analysis of their effects on work-related criteria. *Journal of Applied Psychology*, 84, 496–513.

Blau, P. (1964). *Exchange and Power in Social Life*. New York: Wiley.

Boroff, K., and Lewin, D. (1997). Loyalty, voice, and intent to exit a union firm: A conceptual and empirical analysis. *Industrial and Labor Relations Review*, 51, 50–63.

Bureau of Labor Statistics. (2000). *National Compensation Survey: Survey of Employee Benefits*. Retrieved May 2003, from www.bls.gov.

Clark, S. C. (2000). Work/family border theory: A new theory of work/family balance. *Human Relations*, 53(6), 747–770.

Eaton, S. (2003). If you can use them: Flexibility policies, organizational commitment and perceived performance. *Industrial Relations*, 42, 145–167.

Fedor, D., and Rowland, K. (1989). Investigating supervisor attributions of subordinate performance. *Journal of Management*, 15, 405–416.

Fried, Y., and Ferris, G. R. (1987). The validity of the job characteristics model: A review and meta-analysis. *Personnel Psychology*, 40, 287–322.

Friedman, S. D., Christensen, P., and DeGroot, J. (1998). Work and life: The end of the zero-sum game. *Harvard Business Review*, Nov.–Dec., 119–129.

Golden, L. (2001). Flexible work schedules: Which workers get them? *American Behavioral Scientist*, 44, 1157–1178.

Gottlieb, B., Kellowy, E., and Barham, E. (1988). *Flexible Work Arrangements: Managing the work–family boundary*. Chichester, UK: Wiley.

Grover, S., and Crooker, K. (1995). Who appreciates family-responsive human resource policies: The impact of family-friendly policies on the organizational attachment of parents and non-parents. *Personnel Psychology*, 48, 271–288.

Gutek, B. A., Searle, S., and Klepa, L. (1991). Rational versus gender role explanations for work–family conflict. *Journal of Applied Psychology*, 76, 560–568.

Hackman, R., and Oldham, G. (1980). *Work Redesign*. Reading, MA: Addison-Wesley.

Hill, E. J., Miller, B. C., Weiner, S. P., and Colihan, J. (1998). Influences of the virtual office on aspects of work and work/life balance. *Personnel Psychology*, 5, 667–683.

Homer, L. (2002). Work and home as competing and complementary domains. Symposium presented at the annual meeting of the National Academy of Management, Denver, CO.

Kanter, R. M. (1977). *Work and Family in the United States: A critical review and agenda for research and policy*. New York: Russell Sage Foundation.

Kellogg, K. (2001). From telecommuting to timework: Looking back in order to look ahead. Doctoral paper for graduate course credit, Massachusetts Institute of Technology, Cambridge, MA.

Kossek, E. E. (2006). Work and family in America: Growing tensions between employment policy and a changing workforce. A thirty year perspective. Commissioned chapter by SHRM Foundation and University of California Center for Organizational Effectiveness for the 30th anniversary of the State of Work in America. In E. Lavles and J. O'Toole (Eds.), *America at Work: Choices and challenges* (pp. 53–72). New York: Palgrave Macmillan.

Kossek, E., Barber, A., and Winters, D. (1999). Managerial use of flexible work schedules: The power of peers. *Human Resource Management Journal*, 38, 33–46.

Kossek, E., and Lautsch, B. (2008). *CEO of Me: Creating a life that works in the flexible job age*. New Jersey: Pearson/Wharton Press.

Kossek, E., Lautsch, B., and Eaton, S. (2006). Telecommuting, control, and boundary management: Correlates of policy use and practice, job control, and work–family effectiveness. *Journal of Vocational Behavior*, 68, 347–367.

Kossek, E., Noe, R., and DeMarr, B. (1999). Work–family role synthesis: Individual, family and organizational determinants. *International Journal of Conflict Resolution*, 10, 102–129.

Kossek, E., Roberts, K., Fisher, S., and DeMarr, B. (1998). Career self-management: A quasi-experimental assessment of a training intervention. *Personnel Psychology*, 51, 935–962.

Mokhtarian, P., Bagley, M., and Saloman, I. (1998). The impact of gender, occupation, and presence of children on telecommuting motivations and constraints. *Journal of the American Society for Information Science*, 49, 1115–1134.

Nippert-Eng, C. (1996). *Home and Work: Negotiating boundaries through everyday life*. Chicago: University of Chicago Press.

Perlow, L. (1998). Boundary control: The social ordering of work and family in a high-tech corporation. *Administrative Science Quarterly*, 43(2), 328–357.

Pierce, J. L., Newstorm, J. W., Dunham, R. B., and Barber, A. E. (1989). *Alternative Work Schedules*. Boston: Allyn Bacon.

Poster, W., and Prasad, S. (2005). Work–family relations in transnational perspective: A view from high-tech firms in India and the United States. *Social Problems*, 52(1), 122–146.

Rapoport, R., Bailyn, L., Fletcher, J., and Pruitt, B. (2002). *Beyond Work–Family Balance: Advancing gender equity and workplace performance*. San Francisco: Jossey-Bass.

Rau, B. L., and Hyland, M. (2002). Role conflict and flexible work arrangements: The effects on applicant attraction. *Personnel Psychology*, 55, 111–136.

Roehling, P. V., Roehling, M. V., and Moen, P. (2001). The relationship between work–life policies and practices and employee loyalty: A life course perspective. *Journal of Family and Economic Issues*, 22, 141–170.

Scandura, T., and Lankau, M. (1997). Relationship of gender, family responsibility and flexible work hours to organizational commitment and job satisfaction. *Journal of Organizational Behavior*, 18, 377–391.

Zedeck, S. (1992). Exploring the domain of work and family concerns. In S. Zedeck (Ed.), *Work, Families, and Organizations* (pp. 1–32). San Francisco: Jossey-Bass.

7 Commuting and well-being

Raymond W. Novaco and Oscar I. Gonzalez

The progress of human societies has always been associated with mobility and transportation. Since the earliest homo sapiens migrated from their cave dwellings over 100,000 years ago to establish civilizations in the valleys of the Tigris and Euphrates, human achievement and societal well-being has been dependent on our ability to transport ourselves and the goods that we produce. While the information age has radically modified both work products and the means of transportation, physical mobility on roadway systems remains fundamental to our commerce, recreation, and other life-sustaining activities. Indeed, mobility is very much associated with well-being, but the relationship is not uniform, as higher levels of travel may be reflective of constraints on opportunities, depletion of resources, and impairments to personal health and job performance.

Contemporary urban societies are fraught with stressful environmental conditions, among which traffic congestion features prominently. Remaining attached to the mode of private automobile travel and constrained by the availability of affordable housing, workers endure congested commutes and absorb the stressful consequences. As Dubos (1965, 1969) observed about human adaptation, we seem to develop tolerances to aversive environmental conditions and apparently function effectively in these less than healthy environs; however, as he also asserted, such adaptations in the present will be paid by misery in the future. Although potentially harmful urban conditions, such as exposure to noise, air pollution, and traffic congestion, become acceptable through habituation, the adaptive adjustments are achieved at the price of physical or psychological disturbances later in life. Adaptations to chronic stressors have costs, as was famously established for noise and social stress by Glass and Singer (1972). Because routine exposure to high-density traffic has become normative in urban areas worldwide, attention to the strain of commuting on congested roadways and long-distance commuting is important for both researchers and professionals in health and social sciences, urban planning, engineering, economics, and business management.

In this chapter we will provide an overview of research on commuting stress and its impact on physical health and psychological adjustment. Our focus is on work travel, although we include some relevant stress research on driver performance and the psychological characteristics of drivers found in studies conducted outside of the commuting domain. We will also discuss what has been learned about technologically driven alternatives to cope with commuting stress as it bears on work performance and family relationships as important domains of personal well-being.

At the outset, we acknowledge that there is a diverse array of stress-related research connected with driving and commuting, much of which falls broadly under the rubric of the psychology of transportation – a field of applied psychology that entails the study of human behavior and well-being in conjunction with the regularized movement of people, goods, and services regarding work, personal life, and community activity systems. That general field also involves the interface of humans with transport vehicles and facilities, including their design for optimal operation, provisions for human needs, forms of utilization, road design, and social impacts (Rothengatter and Vaya, 1997). Another related area is driver stress research, which is part of the human performance field of applied psychology.

Our focal domain of commuting has salience in the routines of daily life, and its importance bears on personal, family, and organizational well-being. Because various elements of transport systems are operated by people, understanding the stress-related dimensions has wider relevance for system development, efficiency, and safety – in transit vehicles, as well as automobiles, human cognitive, personality, and performance factors affect vehicle operation. Transport systems shape the structure of our communities, and the experiences we have while using them spill over to home and work domains. People respond to environmental constraints on mobility and strive to enhance well-being. What happens on a particular journey depends on many intertwined psychological elements, including personality disposition, attitudes about the origin and destination of the trip, and resources for choosing alternative travel modes and schedules, including telecommuting. The topic of commuting stress, while centrally focused on health and performance impairments resulting from commute travel impedance, is also inclusive of travel mode decisions, driver attitudes and performance, the interplay between work, family, and emotional well-being, and major societal issues such as the needs of special populations, the definition of the workplace, and the provision of transportation facilities.

Some historical background

The field of transportation has been dominated by engineers, urban planners, and economists; but the contributions of psychologists and sociologists have been integrated into this arena (e.g., Altman *et al.*, 1981; Wachs and Crawford, 1992). Discussions of the social costs of transportation are now common, and psychological elements are given weight in such considerations. Because individuals, families, and work organizations recognize that they absorb hidden costs associated with traffic congestion, air pollution, and traffic noise, Stokols and Novaco (1981) argued that there was considerable merit in understanding how to optimize transportation systems and to assess the consequences of travel constraints.

Involvement by psychologists in the field of transportation began in the 1920s when transport companies and government agencies sought to improve the selection of drivers of public transport vehicles, and psychometric testing became used in choosing the operators of buses, trams, and trains. A central concern was to minimize accident risk, and psychological tests were used to screen applicants who could be anticipated to be accident prone. Wartime studies of adjustment among military personnel and the early psychological formulations of stress, adaptation, and coping that resulted from observations made by psychological researchers working with the military extended quite nicely to the subject of driver performance demands and safety risk. As well, industrial psychology research on fatigue at the workstation transferred to the issue of vigilance in professional train and road drivers (Barjonet, 1997).

As the automobile progressively came to dominate travel in Western societies in the 1950s and 1960s, car ownership grew dramatically; and accident rates and road deaths became serious public health problems. For example, in most European countries, the number of road deaths doubled between 1955 and 1970. Public safety policies, such as speed limits, penalties for drunk driving, and ultimately seat belts, were instituted to curtail casualties, and research into the human causes of accidents came into play. Social psychological knowledge was used in public persuasion efforts, and theories and techniques of learning found fertile ground in driver education and rehabilitation. As the advent of the computer augmented the capacity to analyze and integrate data, the approach in the 1970s and 1980s to understanding driver-based safety risk often took the form of psychological modeling that incorporated personal background, attitude, motivation, and emotion variables, on top of ergonomic factors.

The growth in automobile commuting that occurred between 1960 and 1980, especially in the United States, prompted interest in modeling how trips were generated, how they were distributed across travel zones, and how trips split across available transportation modes (car, bus, train). The boom in automobile commuting followed from substantial growth in the number of jobs, in the working-age population, and in the number of women in the labor force. In addition, homes and jobs shifted to suburban locations. Because neither highway nor transit systems were designed for suburb-to-suburb commuting, traffic congestion in metropolitan areas became a salient issue. In the US, the use of private vehicles between home and work nearly doubled from forty-three million users in 1960 to eighty-three million in 1980. At the same time, alternative modes of travel declined, vehicle occupancy declined, and national investment in highway infrastructure fell far behind the demand for road use (Pisarski, 1987).

European cities, with their treasured historic centers having limited road space and parking facilities, became concerned about the detrimental effects of traffic congestion and pollution and were keen to impose access restrictions. These trends spurred interest in the travel behavior of individuals. Computer-analytic methodology spurred development of travel demand modeling, which blended consumer theory in economics with choice theory in psychology, which assumes that people use some utility or value function in assessing choice alternatives.

Attention to the transportation needs of special populations, such as working mothers, the elderly, and those affected by physical disabilities, also came into focus during the 1970s and 1980s. An important component of the auto-commuting boom was a large increase in the number of women in the labor force (Grieco *et al.*, 1989; Pisarski, 1987). Single women with children who lived in suburbs were very dependent on the automobile (Rosenbloom, 1992). In the US, most trips made by working mothers and their children cannot be made by transit, and the impact of travel constraints on the quality of family life is an important topic. Indeed, Novaco *et al.* (1991) found that high levels of home environment stress were linked to traffic congestion exposure among women in high impedance (long-distance/long-duration) commutes, independent of income and job satisfaction. In the 1990s, telecommuting emerged, partly as a way to reduce strain on family life. The convergence of technological, economic, social, and psychological factors made work less place dependent, and telecommuting was heralded as a viable alternative to auto-commuting (Mokhtarian, 1990).

In the most recent analysis of long-term commuting trends in the US, Pisarski (2006) shows that the booms in suburbanization, in commuting from suburb to suburb, in vehicle ownership, and in commuting by private vehicle that were evident from 1960 through 1980 have continued through the year 2000 census. Half of the US now lives in suburbs, and from 1990 to 2000, 64 percent of the growth in metropolitan commuting was associated with suburb-to-suburb commutes. The next largest growth share (almost 20%) was from central city to suburbs. Such travel patterns intensify congestion, as they are not efficiently accommodated by existing radial road design networks. Strikingly, comparing 1980 with 2000 percentages of modal use share, driving alone increased from 64.4 percent to 75.7 percent, carpooling decreased from 19.7 percent to 12.2 percent, and transit use declined from 6.2 percent to 4.6 percent. Travel time in getting to work is now more than sixty minutes for 8 percent of workers in the US. Increasing numbers of workers are starting their commute before 5 a.m.

Similar trends have occurred in Europe. Since 1970, there has been a 145 percent increase in the transport of people, and 87 percent of that growth has occurred in the road passenger transport sector (European Commission, 2006). The automobile share of passenger travel increased from 74 percent in 1970 to 79 percent in 2003. In comparison, the rail transport share of passenger travel declined from 10 percent to 6 percent in that same period, and the bus/coach share of passenger travel dropped from 13 percent to 8 percent. "The emergence of the information society has not limited the need for mobility – quite the contrary . . . every European citizen travelled 17 kilometers a day in 1970. Today that has risen to 35 km. Against this backdrop, traffic congestion has reached disturbing levels" (European Commission, 2006: 7). Costa *et al.* (1988a), noting the European trend in suburban growth, combined with reduction in the inner cities, called attention to commuting as a stress factor due to its interference with living and working conditions, such as diminished leisure time and work absenteeism.

Population growth in metropolitan areas worldwide clearly has exacerbated traffic congestion as an international problem. Jones's (2002) account of mega-urban regions in Southeast Asia indicates that public transport problems are immense, with travel times to work of two hours being normal, as is the associated lost productivity and poor quality of life. Shimomitsu (1999) reported that 48.2 percent of Japanese workers in major cities spend two or more hours commuting every day (12.7 percent spend three or more hours) and refers to the commuter-train experience as "commuting hell." Studies of stress associated with traffic congestion in Asian cities are sparse, but the Matthews *et al.* (1999)

study with Japanese drivers found some comparability with British studies on "driver stress," although we discuss below some issues in this genre of research. Overall, it is clear that traffic congestion and the strain of commuting through it is a problem worldwide.

Stress, driving, and commuting

There are a variety of domains that might be subsumed under the rubric of commuting stress, including studies concerning driver stress measurement, driver performance and safety, and factors influencing mode choice decisions. In this chapter, we focus on the stress of commuting, as opposed to driving per se that is not done on one's commute to work. For example, driver personality variables or other individual difference factors that are associated with driving behaviors or accident risk are excluded, as are ergonomically oriented studies and driving simulator research. Also outside of our present scope is research on the stress experienced by professional drivers, such as research conducted with urban bus drivers (e.g., Evans, 1994; Evans and Carrere, 1991), long-distance coach drivers (e.g., Raggatt, 1991; Sluiter *et al.*, 1998), truck drivers (e.g., De Croon *et al.*, 2004; Hentschel *et al.*, 1993; Vivoli *et al.*, 1993), and company-car drivers (e.g., Cartwright *et al.*, 1996).

One does find in the "driver stress" literature a research segment concerning violations and accident involvement that is relevant to our commuter stress topic. Such driving behavior indices legitimately constitute stress-related outcomes, and we review findings from studies in this area later. However, because of its bearing on the framing of the commuter stress concept, we note at this juncture that many "driver stress" designated studies come up short on construct validity, because they have been restricted to limited questionnaire assessments. For example, in many studies of "driver stress," the criterion variable has been a single measure, the "Driving Behavior Inventory" (DBI; Gulian *et al.*, 1989). The DBI items are in large proportion about annoyance, irritability, anger, and aggressiveness, and, in that regard, its assessment of stress is truncated. The experience of stress involves much more than an antagonistic sentiment, mood, or inclination. In the affective sphere, in addition to anger-related dispositions, it entails anxiety, depression, and demoralization. Moreover, the environmental factors that induce stress, such as the physical environmental conditions associated with traffic exposure, merit objective assessment.

To properly assess stress as a condition of the human organism, one must get beyond self-report measures; and, when one must rely on self-report procedures, multiple instruments should be utilized. Stress has cognitive, affective, and behavioral disposition dimensions, as well as task performance, physiological reactivity, psychological adjustment, and personal health components. The stress concept has had extensive elaboration since the landmark writings of Selye (1956) and Lazarus (1966), so much so that the three-volume *Encyclopedia of Stress* (Fink, 2000) has now gone to its second edition. A formulation of psychological stress pertinent to the general domain of transportation is given in Stokols and Novaco (1981).

Our principal focus here will be on automobile commuting, although we will give attention to mass transit users. We examine factors bearing on the stress of commuting as it interfaces with the home and work domains, review findings regarding the psychological and physical manifestations of commuting stress, identify moderating variables, and discuss what has been learned about how people cope with the stressful aspects of commuting.

Conceptualizing commuting stress

The topic of commuting stress is given a very fine exposition in the book by Koslowsky *et al.* (1995), who provide an engaging mix of the environmental, psychological, personal health, and organizational factors involved. They offer an elaborate structural model and provide a rich discussion of coping at the individual, organizational, and governmental levels. One springboard for their view of commuting stress is the impedance concept, which has guided our understanding of the subject and is embedded in broader ecological models (Novaco, Stokols, and Milanesi, 1990; Stokols and Novaco, 1981).

Exposure to traffic congestion can be understood as being stressful in terms of the concept of impedance – the blocking or thwarting of movement and goal attainment. The impedance conceptualization of commuting stress was introduced to capture the obstructive, frustrating aspects of road traffic that interfere with behavioral objectives, evoke unpleasant emotional states, and detract from efficient performance and personal satisfaction (Novaco *et al.*, 1979; Stokols and Novaco, 1981; Stokols, Novaco, Stokols, and Campbell, 1978). It was principally operationalized as a three-level factor defined by the conjunction of the distance and duration of the commute, although other objective indices such as road exchanges later served as supplementary indices (Novaco *et al.*, 1990). The greatest impedance would result from traveling long distances slowly, and, conversely, the least impedance occurs when short distances are traveled in small amounts of time. Validation for the impedance concept was first demonstrated in the above studies, which were quasi-experimental investigations involving criterion measures of physiology, task performance, mood states, subjective distress, and physical health problems and involving repeated measures on each day of a week at company sites, plus questionnaires and travel logs. Generally, higher (physical) impedance was associated with higher blood pressure, lower frustration tolerance, more negative mood, more work absences, more colds and flu independent of work absences, and lower residential and job satisfaction. Evans and Carrere (1991), examining a three-level traffic congestion exposure variable with urban bus drivers, reported progressively higher adrenaline and noradrenaline levels and lower perceived job control.

In subsequent studies in our research program, travel impedance was conceptualized as having both physical and perceptual dimensions (Novaco *et al.*, 1990 and 1991). "Physical impedance" can be objectively measured by the distance and time of the journey, along with the number of road exchanges, which represent nodes of congestion. Time of day is also relevant, as congestion is objectively greater in the evening, and commuting stress is compounded by strains from the workday. "Subjective impedance" is the person's perception of commuting constraints, assessed in this research by seventeen questionnaire items. A structured interview could also be done to ascertain this perceived dimension of travel impedance. Subjective impedance overlaps with physical impedance but is not isomorphic with it. Some commuters who have high physical impedance commutes are low in subjectively experienced impedance, whereas others who have low physical impedance routes perceive their commutes to be high in impedance. Subjective impedance was found to be significantly related to evening home mood, health problems, job satisfaction, and residential satisfaction. In testing the conjecture that subjective impedance mediated the relationship between physical impedance and stress outcomes, confirmatory evidence was obtained for a multivariate cluster of personal affect measures but only for some residential satisfaction measures (Novaco *et al.*, 1991).

Because the commute is sandwiched between the home and the job, the demands and affordances of each of these domains, as well as personality factors, bear on the commuting experience. A useful concept in understanding commuting stress is "inter-domain transfer effects," whereby the psychological consequences of environmental conditions in one life domain (home, commuting, work, or recreation) transfer to another, either positively or negatively (Novaco *et al.*, 1990 and 1991). An empirical example of a positive transfer effect is that a high level of residential choice was found to buffer commuting stress. Empirical examples of negative transfer effects are that exposure to commuting stress affected negative mood at home in the evening, controlling for a great many other relevant variables (e.g., income, education, residential choice, months in residence, residential satisfaction, job satisfaction, work social relationships), and that commuting stress was found prospectively to be associated with change in commuters' job location, independent of job satisfaction and residential location and satisfaction (Novaco *et al.*, 1990 and 1991).

The impedance findings from the measurement-intensive quasi-experimental studies in field settings were extended in a large survey study by Novaco and Collier (1994), conducted with a representative sample of southern California commuters. Commuting stress was found to be significantly associated with distance and duration of the commute, controlling for age and income. Independent validation for the impedance conception of commuting stress occurred in research by Koslowsky and his colleagues. In a survey study with nurses at several hospitals, Koslowsky and Krausz (1993) found that automobile drivers' employee stress was linked to commuting time (they did not assess distance) and that stress mediated employee job satisfaction, commitment, and intention to leave. Unfortunately, they did not assess whether their measure of employee stress, a twenty-item self-report questionnaire, was related to individuals' perception of their commutes. Subsequently, Koslowsky *et al.* (1996), in a survey study of commuters in Tel Aviv, operationalized commute impedance as a three-level physical dimension and found it to be significantly related to subjective stress. They also found that higher levels of commuting impedance were associated with much lower levels of perceived control.

Perceptions of control and predictability were important themes in the nascence of the stress field (e.g., Averill, 1973; Folkman, 1984; Glass and Singer, 1972; Lefcourt, 1973) and remain so (e.g., Bollini *et al.*, 2004; Frazier *et al.*, 2004; Troup and Dewe, 2002). In the commuting stress domain, these themes have also emerged, as has a related concept of commute variability (Kluger, 1998). Environmental controllability is a core property of the Stokols and Novaco (1981) ecological model of commute travel impedance, as it bears on both exposure to commute travel stressors and the person's level of need facilitation in the commuting context vis-à-vis expected or preferred outcomes.

The relevance of perceived control for commuting stress surfaced in the Singer *et al.* (1977) study of train commuters, where it was found that perceived control (associated with getting on the train at the start of the transit line and getting a desired seat) was inversely related to psychophysiological stress (urinary secretion of catecholamines). For automobile commuters, Novaco *et al.* (1979) examined personal control expectancies ("locus of control") as a personality variable and situational control dimensions (residential choice and type of car), finding strongest support for the situational control variables on task performance (frustration tolerance in puzzle tasks and digit symbol recall). The control theme for automobile commuters was pursued by Schaeffer *et al.* (1988), as they investigated differences in blood pressure, heart rate, mood, and task performance between drivers as a function of degree of route choice and whether the person drove alone or was a carpool driver. Having route choice did not buffer stress, and the results for being a solo driver (car environment control) were mixed. Carpool drivers had significantly higher blood pressure and heart rates, but solo drivers had higher hostility and anxiety scores. Also in that study, impedance, indexed by average speed, was significantly related to systolic blood pressure and to proofreading errors. Novaco *et al.* (1990) found a set of choice variables (concerning car selection, commuting route, and residential selection) are significantly related to perceived commuting impedance – the higher the degree of choice, the lower the subjective impedance.

Evans and Carrere (1991) deftly uncovered the stress-mitigating effects of perceived control in their urban bus driver study, showing that perceptions of control (a four-item measure concerning work decisional latitude, control, and freedom) were significantly related to adrenaline and noradrenaline, after accounting for exposure to traffic congestion. As they found that perceived control decreased with higher levels of traffic congestion, Evans and Carrere conjectured that congestion curtails drivers' ability to adjust driving speeds to stay on schedule and to maneuver the bus to pick up and drop off passengers.

In this context, Koslowsky *et al.* (1995) put forward the "predictability" of the commute as an important moderator between subjective impedance and stress, preferring this concept to "control." For example, while having route choice might be thought to reflect control, they did not find it to moderate the number of stress symptoms reported – this is consistent with what occurred in the high-impedance condition in Schaeffer *et al.* (1988). Koslowsky *et al.* asserted that having route choices will not amount to having control when those choices are low in quality. "The commuter put into a situation of choosing between undesirable alternatives may indeed experience more, not less stress" (1995: 100). Thus, Koslowsky and his colleagues proposed predictability as the hypothetical moderator. For example, a person who can simply predict the duration of his or her commute may thereby have the perception of control. Kluger (1998) sought to sharpen this further by proffering the concept of "commute variability." When a commute has low variability across days, the aversiveness of the commute is predictable and uncertainty is low, hence the commuter can maintain a sense of control. Variable commutes, even if predictable, are thought by Kluger to limit the cognitive ability of the commuter to control the environment. Testing this commute variability concept in conjunction with the impedance model, Kluger's study, involving a questionnaire given to graduate-student commuters, found support for the impedance model – commute length was related to perceived commuting strain and to somatic and affective symptoms. While commute choice had some effect, variability (a composite of three indices) had a much stronger effect in relationship to the strain and symptom measures. Variability had the highest correlation with perceived commuting strain. Both commute length and variability had significant partial correlations with symptoms, controlling for perceived commuting strain.

Building on these ideas concerning predictability and variability of Koslowsky *et al.* (1995) and Kluger (1998), Evans *et al.* (2002) investigated whether commute predictability for train passengers affected perceived stress of the commute, salivary cortisol, and performance on a proof-reading task. They trichotomized predictability, which was indexed by a five-item scale, and found that with increasing levels of unpredictability, perceived stress increased, as did salivary cortisol. Proof-reading was unaffected.

Taken together, this body of research points to the scientific value of understanding commuting stress in terms of impedance and that predictability or variability is an important moderator. The best way to index impedance remains to be ascertained, and not all studies that have employed that concept have fully operationalized its physical and subjective dimensions. In some cases, what Novaco *et al.* (1990 and 1991) have termed subjective impedance is treated in its operational forms by other investigators as a stress measure, thus confusing independent with dependent variables – that is, treating antecedent conditions as outcome criteria. As well, the concept of control – whether in the form of choice, predictability, or variability – has been shown in various studies to have empirical and theoretical value. Costa *et al.* (1988b) noted that uncertainty of travel time was often a source of anxiety for their large sample of long-distance commuters in various travel modes. The control concept has also not been sufficiently operationalized, particularly when one considers the ecological context of commuting and the relevance of the home and work domains.

Another potential moderator is time urgency as a *personality* factor (e.g., Stokols, Novaco, Stokols, and Campbell, 1978), which would seem to merit further examination than it has received. Feeling "rushed" in getting to work served to distinguish the highly stressed long-distance females in Novaco *et al.* (1991), and being "in a hurry" was found by Gulian *et al.* (1990) to be very significantly related to multiple stress indices. In a fuller analysis, Hennessy and Wiesenthal (1999) and Hennessy *et al.* (2000) found that time urgency (operationalized by a three-item rating scale) was significantly related to driver stress assessed on time in the vehicle (see procedure described below). Although their time urgency index was not a personality measure, their results lend support to the prospective merit of examining this factor more fully, so as to sharpen conceptions of commuting stress. The suggestion here is for investigators to consider time urgency as a personal disposition variable that affects how commuting experiences, including the anticipation of stressors on destination, are processed by the individual.

Adverse impacts of commuting stress: additional moderators

The various quasi-experimental and survey studies discussed above have shown that commuting stress is associated with elevated physiological arousal, negative mood on arrival at work, negative mood at home in the evening, lowered frustration tolerance, cognitive-performance impairments, illness occasions, work absences, job instability, lowered residential satisfaction, and decrements in overall life satisfaction. As it is easily argued that such investigations suffer from differential selectivity in sample composition (i.e., self-selection into commuting conditions), one must be cautious in suggesting causal connections between commuting strains and stress outcomes. The problem of course for an experimental investigation is that it is not feasible to randomly assign people to commutes of varying degrees of impedance and have them endure those journeys as routine. However, an experimental study by White and Rotton (1998) did randomly assign participants to conditions (either car, bus, or control) for a single commute (travel between two campuses) and obtained stress measures. Those who drove had a significantly higher blood pressure change than the other two conditions and a significantly greater increase in heart rate and lower frustration tolerance compared to controls. The findings of White and Rotton (1998), while limited in ecological validity – students in an experiment sent on a journey between two campuses, as opposed to workers on daily commutes – do provide support for the various field research studies finding physiological (cardiovascular and hormonal) and task performance effects for commuting stress.

At this juncture, we must mention that gender seems to be an important moderator of commuting stress. Novaco *et al.* (1991) found that female solo drivers in high-impedance commutes were the most negatively affected commuters, independent of job involvement and household income. Because women tend to have more logistical responsibilities for home and family, they are perhaps more sensitized to stress on high-impedance routes. The gender moderation effect (females showing higher stress) was also obtained by Novaco and Collier (1994), as women in the long-distance commutes (greater than 20 miles to work) perceived greater commuting stress spillover to work and to home. Koslowsky *et al.* (1996) found a near-significant moderation effect for gender on subjective stress, but they only report the main effect testing, not in interaction with impedance. However, they did find that women commuters had significantly lower levels of perceived control. In studies with rail commuters, Evans *et al.* (2002) and Evans and Wener (2006) found that gender did not interact with commute variables in relationship to stress measures. However, the ecology of commuting by train is quite different from that by automobile, including margins of departure and arrival, intermediate stops, and tasks that can be done during the commute. Hennessy *et al.* (2000) found no gender differences in driver stress, but their participants were on single trips on their routine commutes. In the large sample survey study of Italian factory workers by Costa *et al.* (1988b), women across commuting modes had significantly shorter-duration commutes than did men, but they had significantly higher psychological stress, family problems, illnesses, work absences, and lower job satisfaction. Those authors concluded that (long) commutes accentuate the problems that working women face.

In our preceding discussion, we have given some attention to commuting stress involving public transport, although automobile commuting features more prominently. Comparative analyses of modes of commuting on stress dimensions are sparse, particularly if one looks for measurement intensity. A recent study by Gatersleben and Uzzell (2007) compared drivers, cyclists, walkers, and public transport users on questionnaire measures, and their discriminant analysis shows that car drivers have the most stressful commutes, but public transport users had the most negative attitudes toward their travel mode. Parallel to automobile commuting studies, Evans and Wener (2006) found that the duration of the commute is related to salivary cortisol elevation and proof-reading errors. Unfortunately, the latter study did not track the predictability or control dimension of earlier train passenger studies (e.g., Evans *et al.*, 2002; Singer *et al.*, 1977). The large sample survey study of Italian factory workers by Costa *et al.* (1988b) examined differences between commuters who have long-duration (forty-five minutes or more) versus short-duration (twenty minutes or less) journeys to work – they oddly termed the former "commuters" and the latter "non-commuters," but did not disentangle modes within these conditions. The long-distance commuters (especially the females) did, however, report significantly more discomfort, distress, and somatic complaints, as well as a higher rate of illnesses and sickness absences. McLennan and Bennetts (2003) found that reported stress and productivity decrements varied with length of journey and travel mode. For those in long journeys, car drivers reported more stress, whereas, in short journeys, public transit users were more stressed.

A commute occurs in two directions, but in the vast majority of studies the measures obtained have been in conjunction with the morning commute. In the impedance studies conducted by Novaco *et al.* (1990, 1991) we separately assessed morning and evening commute dimensions. The evening commute is objectively more stressful than the morning commute; it is subjectively perceived as more stressful; and the relationship with residential domain stress outcome measures is stronger for the evening commute impedance than that for the morning. The results of the diary study by Gulian *et al.* (1990) converge in this regard. Daily driving stress scores were significantly higher across days of the week. The stress reported in daily diaries was significantly related to health complaints, relationships at work, and cognitive failures. Gulian *et al.* also found that driving stress was inversely related to age and driving experience.

Pertinent to moderator variable analyses are ideas introduced by Hennessy and Wiesenthal (1997, 1999; Hennessy *et al.*, 2000) concerning state and trait conception of driver stress. They developed a 26-item State Driver Stress Inventory and a State Coping Behavior Inventory relevant to driving (Hennessy and Wiesenthal, 1997), involving a small sample of student and community drivers in York, Canada, who made telephone calls to the investigators at journey landmark points, at which time the state stress and coping measures were administered. Trait stress scores were computed from a general driving behavior inventory, and trait coping behaviors from another separately administered inventory. The drivers commuted on either low- or high-congestion routes. The state driver stress varied significantly as a function of both trait driver stress and level of congestion. Hennessy and Wiesenthal (1999) extended this line of research to aggressive behavior, using a larger sample in the same region. Driver aggressive behavior, assessed verbally by the telephone call procedure, was predictive of state driver stress, especially in the high-congestion condition. Subsequently, Hennessy *et al.* (2000) replicated their previous findings that state driver stress is a function of level of traffic congestion, time urgency, and trait susceptibility toward driver stress. They also found that daily hassles were carried forward into the driving environment, amplifying driver stress levels among high trait stress drivers. This lends support to the notion of "inter-domain transfer effects" (Novaco *et al.*, 1990), with results obtained from online assessments in the vehicle. While this research by Hennessy and Wiesenthal has been conducted on single trips, these did occur on the person's routine commute route. Their use of a state/trait distinction for driver stress and their innovative measurement of the state variables offers promise for future research discoveries.

Driver stress, violations, and accidents

Beyond the health and psychological indicators of stress reflected in the studies we have heretofore discussed, psychological stress variables affect the operation of transport vehicles, as can be seen in studies of driver error and performance decrements (Rothengatter and Vaya, 1997). Fatigue has been the most researched topic in this regard. Ordinarily understood as physical exhaustion, it can also involve conditions characterized by boredom, task aversion, uncontrollable sleepiness, hypnotic trance-like states, or loss of alertness. Driving performance decrements occur before the driver is subjectively aware of the fatigue. Fatigued drivers become complacent and underestimate the effort needed to maintain driving performance. This issue is of course most applicable to professional drivers, for whom limitations on hours of service are implemented to guard against fatigue-related safety errors, but it also has relevance to stress spillover from the workday on the commute home. However, we here turn to driving violations and accidents as additional stress criteria.

"Driver stress" has received considerable research attention as a focal topic (e.g., Hartley and El Hassani, 1994; Kontogiannis, 2006; Simon and Corbett, 1996; Westerman and Haigney, 2000). Studies have typically included one, or both, of two self-report measures: the Driving Behavior Inventory (DBI; Gulian *et al.*, 1989) and the Driving Behavior Questionnaire (DBQ; Parker *et al.*, 1995; Reason *et al.*, 1990). Despite the extent of their use, their construct validity is not so well established. For example, most of the DBI items are not about behaviors, but rather concern emotional reactions and cognitions. There is a driver stress subscale within the DBI measure, but this scale is heavily loaded with items having anger/frustration/aggression content. In contrast, the Hennessy and Wiesenthal (1997) state driver stress measure samples a much wider range of affects. Moreover, many of the dimensions of human stress reflected in the studies discussed above, such as dysphoric mood, cognitive interference, dissatisfaction in various life domains, and health complaints, are not incorporated in the DBI. Nevertheless, these instruments have had extensive use, and the research involved merits our attention.

Kontogiannis (2006) proposed a model of accident causation in which distal factors (e.g., driver stress) create a disposition to exhibit risky driving behaviors (e.g., errors and violations), which in turn predict actual accident involvement. High driver stress, measured by the DBI, was associated with higher levels of errors and violations measured by the DBQ. However, item inspection of the DBI and DBQ draws out two serious caveats. First, as we indicated, the DBI captures not only driver stress (a state of physical or emotional strain), but also driver aggression (driving behaviors). Second, both instruments share content related to driver aggression. The fact that both distal and proximal factors measure aggression is methodologically muddled.

The DBQ (Reason *et al.*, 1990) has been repeatedly translated, factor analyzed, and applied in a range of cultural settings. The study that brought forth the DBQ aimed at investigating the probable differentiation between errors and violations while driving. A sample of drivers from the UK anonymously completed and returned the questionnaire by mail. Three distinctive factors were identified: violations, dangerous errors, and relatively harmless lapses. Thereafter, the study was replicated in a Western Australian sample (Blockley and Hartley, 1995) and in a Swedish sample (Aberg and Rimmo, 1998) yielding somewhat similar results.

There have been moderately consistent findings for the DBI. For demographic variables, the number of driving violations decreases with age; regarding gender, males report more violations, traffic accident involvement (as drivers), license suspension, and convictions for speeding than do females (Aberg and Rimmo, 1998; Blockley and Hartley, 1995; Reason *et al.*, 1990). Not unexpectedly, roadway exposure, measured in hours of driving and distance driven per week, is a significant predictor of dangerous violations (Blockley and Hartley, 1995; Reason *et al.*, 1990).

Consistent with the concept of interdomain transfer effects, stressful life circumstances outside the context of driving are associated with errant driving behaviors. Veneziano *et al.* (1994) found that driving while intoxicated (DWI) arrestees were more likely to report stressful life events (e.g., job loss and unemployment, financial difficulties, divorce and separation), and being in an accident, than would be expected in the general population. Nearly a quarter of the sample (23.9%) indicated traffic accident involvement within the last year. The cross-sectional design of this study does not entail directionality, so our ability to draw inferences of causality is restricted. An increase in stressful circumstances could lead to alcohol abuse, or alternatively, increased drinking can lead to stressful circumstances.

Getting beyond cross-sectional design limitations, Lagarde and colleagues (2004) conducted a prospective investigation over an eight-year period, exploring the association between stressful life events, namely marital separation and divorce, and traffic accidents in a large cohort of French workers. Consistent with previous findings, those with higher annual driving mileage – men and younger drivers – reported more accidents. Remarkably, separation or divorce was strongly associated with higher risk of serious accidents and even more strongly with at-fault status. Other potentially stressful life events, such as a child leaving home and an important purchase, were also associated with serious traffic accident involvement.

Corroborating evidence supports the notion that negative emotions and stressful life events can adversely affect driver performance. Negative mood has been found to predict violations, hazardous errors, and nonhazardous errors (Reason *et al.*, 1990). Similarly, Legree *et al.* (2003) found that the negative emotional state preceding an accident (e.g., stressed versus calmed, stressed due to life event, fatigue, sleep deprivation, being disturbed by a passenger) predicted at-fault status. When combined, results of these studies suggest that elevated stress while driving is a risk factor that increases the likelihood of driving errors, violations, and traffic accident involvement.

Environmental conditions affecting driver stress and coping

Although stress has been defined in many ways, at its core stress is grounded on the relationship between the person and the environment (Lazarus and Folkman, 1984). Research on how stressors (environmental demands) affect perception and information processing during tasks, including driving, is not new (Brown, 1962; Brown *et al.*, 1966; Finkelman and Glass, 1970; Plutchik, 1959). Given that excess stress and strain can hinder driving performance and increase accident likelihood during a commute, the study of stressors in the roadway environment is obviously important. Only a fraction of the numerous environmental stressors present while driving have been systematically researched.

Finkelman *et al.* (1977), examining the effect of noise on driver performance, hypothesized that drivers allocate their perceptual and information-processing capability to the demands of the immediate driving tasks and use the remaining capacity on subsidiary tasks. Their study assessed the demands of separate and conjoint task (driving and delayed digit recall) coupled with environmental stress (nonperiodic noise) on drivers' perceptual and processing capacity. Nonperiodic noise significantly increased driving task time, but only degraded performance under the highest load condition (conjoint task). A remarkable finding was that, in this condition, drivers were much more likely to reduce accuracy than speed. We are inclined to mention that small rubber pylons, used to mark driver accuracy, may not attract the same amount of attention as real-world markers (e.g., center lane dividers and guard rails).

Like noise, heat is a potentially aversive environmental element. Wyon *et al.* (1996) investigated whether moderate heat stress (21° or 27°C compartment temperature) reduced individuals' vigilance for a set of onboard signals while driving in actual traffic in Sweden. Average response times, across all signals, were significantly increased in the higher heat condition. Interestingly, data analysis revealed no significant differences until the second half hour of the driving task. Six of the eighteen female subjects in the high-heat condition erred by a lapse of attention, while no such errors were recorded for females in the low-heat condition. The manipulation of cabin temperature was not confounded by external weather, and none of the subjects guessed that thermal conditions were under investigation. Consistent with previous research (Bursill, 1958; Mackie *et al.*, 1974), findings suggest that even moderate increases in compartment temperature, insufficient to cause observable sweating in all but a few subjects, could adversely affect driver vigilance.

Hill and Boyle (2007) recently explored how different tasks and roadway conditions influence the amount of stress perceived by drivers. A factor analysis of eighteen driving scenarios from a nationally distributed survey yielded four factors. The first factor was weather-related stress (e.g., heavy rain, snow). The second factor involved stress resulting from interactions with other drivers (e.g., driving behind a vehicle that is constantly braking). The third factor accounted for stress related to performing driving tasks (e.g., merge into heavy traffic). Finally, the fourth factor accounted for stress from conditions related to limited visibility (e.g., night driving). Analyses showed that females were much more likely to be stressed by adverse weather conditions, and other drivers. Increases in age were associated with increases in driver stress. A significant interaction effect between age and gender revealed that, as age increased among women, lower stress levels were reported when compared with men. Females, older drivers, and drivers with more crashes reported higher levels of visibility-related stress. No significant differences were found when taking into account freeway versus nonfreeway driving.

Contrasting the previously reviewed studies, a smaller amount of research has explored factors in the roadway environment that might potentially reduce stress and/or ameliorate well-being. Parsons, Tassinary, Ulrich, Hebl, and Grossman-Alexander (1998) examined whether stress recovery and/or immunization varied as a function of the roadside environment. This was attempted by showing participants one of four different video-taped simulated drives, varying in natural vegetation and appeal, immediately after and preceding mildly stressful events (passive or active). Stress responses were captured by a number of physiological measures including: electrocardiogram (ECG), skin conductance, facial electromyogram (EMG), and electrooculographic (EOG) signals. There was some evidence suggesting that exposure to artifact-dominated roadside environments both slowed down and impeded recovery from laboratory stressors relative to exposure to nature-dominated roadside environments. In contrast, exposure to the nature-dominated condition decreased the magnitude of the autonomic response to the lab stressors and concomitantly increased coping ability, as reflected in task performance.

More recently, Cackowski and Nasar (2003) similarly explored the potential mitigating effects of highway vegetation on drivers' anger and frustration. After exposure to a ten-minute stressor (a timed addition task with random noise), participants were given the State-Trait Anger Expression Inventory (STAXI; Spielberger, 1996). They then randomly viewed one of three video tapes of highway drives, varying in the amount of vegetation and man-made material, followed by a second stress-inducing task (timed anagrams, some unsolvable). The time spent on the unsolvable anagrams served as a measure of frustration – more frustration inferred from less time spent. Unfortunately, anger dissipated substantially across conditions during the short video, precluding a test of changes across roadside conditions. However, results for frustration tolerance supported the hypothesis: respondents viewing roadways with more vegetation worked almost a minute and a half longer than respondents who viewed other videos, consistent with previous findings (Parsons *et al.*, 1998).

An environmental variable naturalistically associated with driving and driver efforts to cope with stress is music. Wiesenthal *et al.* (2000) examined the efficacy of music in dealing with stress while driving by randomly assigning commuters to either a music or nonmusic group. State measures of driver stress (Hennessy and Wiesenthal, 1997) were obtained via their cellular telephone procedure, during a single commute in low- and high-congestion conditions. A significant interaction was obtained between congestion level and music groups. As in their other studies discussed above, greater stress was reported in high congestion, with the highest stress in the nonmusic group. This finding supports the notion that music could have stress-reducing effects in high-congestion commutes.

The hypothesized benefits of reducing stress while driving range from increased processing capacity and attentiveness to reduced negative emotional states. Hypothetically, stress-ameliorating environmental factors might indirectly reduce traffic accident incidence. Relatively little attention has been given to stress-mitigating roadway/driving conditions or to commuters' efforts to cope with the stress experienced while driving. However, one of the ways of coping with commuting stress is to avoid the commute itself, and technological innovations that have moved the workplace have made this possible.

Commuting as enhancement to well-being

The view that commuting is an unhealthy, derived demand (Mokhtarian and Salomon, 2001; Mokhtarian *et al.*, 2001) has been the subject of much empirical work, as our discussion of commuting stress has documented. However, commuting is not intrinsically harmful nor is it a uniformly unwanted activity pattern. The popular notion of our "love affair with cars" (cf. Marsh and Collett, 1986; Wachs and Crawford, 1992) has corroboration in Steg's (2005) factor-analytic identification of symbolic and affective motives for car use (along with instrumental ones). Also, qualitative studies suggest that commuting might be understood as a gift rather than as a curse (e.g., Jain and Lyons, 2008; Mann and Abraham, 2006).

Despite the widespread increase in traffic congestion and the associated intensification of travel impedance that we have portrayed, most individuals do want some commuting time, many are relatively satisfied with their commute, and a few actually prefer an increase (Redmond and Mokhtarian, 2001). Many commuters may very well choose to drive more than is needed, and few individuals view car ownership as a burden (Handy *et al.*, 2005). Commuting is indeed seen as having positive utility by both automobile drivers and transit users, as shown in multiple studies by Mokhtarian and her colleagues (Mokhtarian *et al.*, 2001; Ory and Mokhtarian, 2005; Redmond and Mokhtarian, 2001), identifying a plethora of utility dimensions, from mobility to lifestyle. Beyond the obvious utility of reaching a destination, people find utility in the commuting domain for doing work tasks, social networking, personal isolation/relaxation, and daily activity planning. Travel time has many information-age utilities, as suggested by Lyons and Chatterjee (2008) and Lyons and Urry (2005).

The affinity for automobile commuting entails more than utilities. For example, when making choices about vehicle type, destination, route, and risk-taking, commuters often willfully disregard objective, instrumental considerations. While a wide range of objective variables (e.g., trip frequency, distance, duration, destination, and vehicle type) influence commuters' judgements of travel, various person-determined factors (e.g., attitudes and personality) are also operative (Collantes and Mokhtarian, 2007). Some drivers want and enjoy commuting, as illustrated by the qualitative studies by Gardner and Abraham (2007) and Mann and Abraham (2006), and private automobile transport is endorsed by commuters as conveying greater psychosocial benefits (e.g., mastery, self-esteem, protection, autonomy, and prestige) than does public transport (Ellaway *et al.*, 2003).

It has long been considered that the psychological reasons for car use (e.g., emotional attachment, habit formation, prestige) outweigh the practical ones (e.g., Stokes and Hallett, 1992). Research by Steg and colleagues (Steg, 2005; Steg *et al.*, 2001) has found that car use is more strongly related to symbolic (status and prestige) and affective motives (freedom, fun, love of driving) than to instrumental ones (transportation characteristics) and that drivers make a clear distinction between these categories.

While investigation of the potential psychosocial benefits of commuting is in its nascence, we suggest that studies in this regard get beyond the all-too-easy self-report questionnaire mode. The physiological, work performance, and physical health aspects should be examined. As transport policies sometimes aim at reducing vehicular use to increase public health, physical well-being criteria should be utilized. Epidemiological research by Macintyre *et al.* (2001) found that car access/ownership is associated with better physical health, controlling for age and socio-demographic variables (social class, marital status, housing tenure, and income). Given that other studies (e.g., Hiscock *et al.*, 2002) have found psychosocial benefits associated with car use, the interrelationship of the psychological and physical health benefits, and the mechanisms involved, merit examination.

Telecommuting and well-being

Telecommuting, a term coined by Nilles (1975) in the early 1970s, refers to "travel" to the office by way of telecommunications. Recent estimates indicate that the number of telecommuters has risen to over twenty-eight million in the United States (WorldatWork, 2007). This figure reflects a growing recognition of communications, especially computer technology, as an existing alternative capable of significantly reducing commutes.

Motives for telecommuting are bountiful, but, since telecommuting leads to commute reduction, it is not surprising that many telecommuters find value in this result (Hartig *et al.*, 2007; Stephens and Szajna, 1998). Stephens and Szajna (1998) found evidence confirming that commute reduction implies less driver frustration and stress, as conjectured by Trent, Smith, and Wood (1994). Other widely reported motivations for telecommuting are: flexibility, relaxed atmosphere, cost savings, increased family time, and avoidance of office politics. Telecommuting enables workers to maximize control over their work schedules and transportation modes; and an abundance of research on human stress in the past two decades has shown that personal control mitigates stress. Such beneficial effects of control have been found with regard to noise, crowding, air pollution, institutionalization, surgery, and stressful life events (Cohen *et al.*, 1986, as well as in the commuting studies previously discussed). A number of reviews concerning telecommuting, motivations, and psychological well-being have been published in recent years (e.g., Andriessen, 1991; Baruch, 2001; Konradt *et al.*, 2000). As motives and circumstances for telework vary, telecommuting can affect well-being in different ways.

The relationship between telecommunications and transportation can be seen to be symbiotic (Mokhtarian, 1990), and it is quite possible that advances in telecommunications technologies will result in travel increases. As Mokhtarian states, the most important impact of telecommunications is in the reduction of constraints. At the psychological level, this has important implications for the well-being of workers, consistent with the broader issue of "personal control." Within most work organizations, issues of personal control are recurrent themes (Greenberger *et al.*, 1991); workers generally seek to enhance control, and traditional management routinely attempts to maintain control over subordinates. Kossek *et al.* (2006) reported that employees who perceived greater job control (the degree to which an individual perceives control over where, when, and how work is done) had significantly lower turnover intentions, family–work conflict, and depression. Kossek *et al.* expand on the personal control and work flexibility theme in Chapter 6.

Although management preference for line-of-sight supervision is a commonly noted institutional impediment to telecommuting, it is noteworthy that not everyone capable of telecommuting chooses to do so. An inspection of the literature reveals numerous reasons why telecommuting can be disadvantageous, including: decreased social interaction, lack of career visibility, the need for direct supervision, and a desire to separate home and work life. Standen *et al.* (1999) suggested that the erosion of boundaries (the structural and psychological divide between life and work) negatively affects well-being. Much attention has also been given to the potential consequences of overlap between work and nonwork domains (Ellison, 1999; Hill *et al.*, 1996), where a home becomes negatively isolating (Ellison, 1999; Lundberg and Lindfors, 2002) and individuals perceive less social support (Trent *et al.*, 1994). Some have contemplated how telecommuting may undermine restorative functions of the home (Hartig *et al.*, 2007) and considered how family-related stressors can spill over to the work domain (Baruch and Nicholson, 1997; Lewis and Cooper, 1995; Standen, 2000).

Telecommuting is viewed as a coping strategy to enhance congruence between demands and resources. However, only a handful of studies have examined personal outcomes of telecommuting by comparing telecommuters versus commuters, and the results are inconclusive. Belanger (1999) found that telecommuters report greater productivity and personal control than commuters but obtained no significant differences in either satisfaction or performance, whereas Kossek *et al.* (2006) found telecommuters had higher performance ratings by supervisors (although that did not result from flexibility control; see Chapter 6). As periodic telecommuters and regular office workers have been found to have higher rates of social support than full telecommuters (Trent *et al.*, 1994), the relationship between telecommuting and social isolation requires examination. Golden (2006) did find a higher degree of teleworking to be associated with more organizational commitment and less turnover intentions, with work exhaustion partially mediating the relationship between degree of telework and organizational commitment and fully mediating the relationship between telework and turnover intentions.

Concerning psychological well-being and the family home environment, in Kossek *et al.* (2006) telecommuting was related to higher depression, except for women with children, who had lower depression. Not surprisingly, the amount of hours of telework was positively related to work-to-family conflict. They asserted that "all forms of flexibility are not necessarily good for individual well-being" (2006: 363). In the Swedish study by Hartig *et al.* (2007), teleworkers and nonteleworkers both experienced the home as a place of restoration far more than as a place of demands. However, as overlap became higher, the teleworkers' evaluation of the home shifted from positive to negative. Working at home facilitates interdomain transfer effects, and the outcomes are not straightforward. That can be seen in Golden *et al.*'s (2006) results that the more extensively individuals telecommuted, the less their work interfered with their family (WFC), but the more their family interfered with their work (FWC). Job autonomy and schedule flexibility positively moderated telecommuting's impact on WFC, but household size was found to negatively moderate telecommuting's impact on FWC. In a similar vein, Konradt *et al.* (2003) found that home-centered telecommuters reported significantly lower task-related stressors (interruptions, concentration demands, time pressure, and uncertainty) but higher nonjob-related stressors (strains due to restricted material resources, burdened duties at home, and social factors) than office-centered telecommuters and regular office workers. Kossek *et al.* (Chapter 6) expand on the boundary-management issue (see their excerpts of participants' reports regarding family and friends, p. 163).

Forecasts of the diffusion of the telecommuting innovation have projected significant benefits in personal well-being and in reductions in traffic congestion, but outcomes are far from clear. The personal privacy and control afforded by the automobile, as well as its embeddedness in Western societies, suggest that it will be quite difficult to dislodge solo drivers from their private cars without imposing economic penalties or governmental regulations.

In addition to telecommuting, the information age had many effects on transportation systems. The development of intelligent transport systems, such as in-vehicle devices to aid navigation (e.g., route guidance and congestion warnings), as well as the increased sophistication of driver panel displays, raised the importance of driver control, visual search efficiency, and mental load on driving proficiency. On the horizon of advanced technologies are Intelligent Vehicle Highway Systems that automatically control steering, speed, and spacing. However, having a vehicle travel under the control of a computerized system raises psychological issues of personal control and freedom, which have been themes intrinsic to car ownership and driving.

Adaptation: an overarching concept

Dissatisfaction with the commute is a cost the employee absorbs, often with delayed effects on health and family life, as well as on work organizations – commuting stress can be manifested in illness-related absences from work and with impaired work performance. People will change jobs, commuting modes, vehicles, or commuting schedule to cope with traffic congestion, but such adaptations are a function of coping resources.

A core concept in the psychology of transportation is behavioral adaptation. Drivers and transport users will adjust to conditions in the transportation environment, whether that be in their overall commuting scheme, roadway or transit conditions, vehicle conditions, or changes in skill level or state of functioning in order to enhance their well-being. Thus, adaptation occurs at different levels of analysis, ranging from "high-level" trip and travel mode decisions to "low-level" vehicle control behaviors that maintain safety margins. When behavioral adjustments involved in adaptation are unsuccessful in the long term, more substantial behavioral coping can be expected, as when someone changes jobs to reduce commuting strain or stops driving because of age-related driving skill decrements. Innovations in information technology, by decreasing the need for vehicle trips, progressively may facilitate adaptation to travel constraints imposed by congested roadways and to unwanted reductions in mobility for some populations. The commuting stress model presented in Novaco *et al.* (1990) incorporates the transportation environment, commuting stress (physical and subjective impedance), and cognitive and behavioral coping, as well as stress and resource factors in the home, commuting, and work domains. The interdomain and longitudinal complexities remain to be empirically examined.

We should stay mindful of the wisdom of Rene Dubos:

> Admittedly, human beings are so adaptable that they can survive, function, and multiply despite malnutrition, environmental pollution, excessive sensory stimulation, ugliness and boredom, high population density and its attendant regimentation. But while biological adaptation is an asset for the survival of *Homo sapiens* considered as a biological species, it can cause a progressive loss of the attributes that account for the peculiarities of human life. (Dubos, 1969: 80)

References

Aberg, L., and Rimmo, P. A. (1998). Dimensions of aberrant driver behaviour. *Ergonomics*, 41, 39–56.

Altman, I., Wohlwill, J. F., and Everett, P. B. (1981). *Transportation and Behavior*. New York: Plenum Press.

Andriessen, J. H. E. (1991). Mediated communication and new organizational forms. *International Review of Industrial and Organizational Psychology*, 6, 18–70.

Averill, J. (1973). Personal control over aversive stimulation and its relationship to stress. *Psychological Bulletin*, 80, 286–303.

Barjonet, P. (1997). Transport psychology in Europe: A general overview. In T. Rothengatter and E. C. Vaya (Eds.), *Traffic and Transport Psychology: Theory and application* (pp. 21–30). New York: Pergamon.

Baruch, Y. (2001). The status of research on teleworking and an agenda for future research. *International Journal of Management Reviews*, 3, 113–129.

Baruch, Y., and Nicholson, N. (1997). Home, sweet work: Requirements for effective home-working. *Journal of General Management*, 23, 15–30.

Belanger, F. (1999). Workers' propensity to telecommute: An empirical study. *Information & Management*, 35, 139–153.

Blockley, P. N., and Hartley, L. R. (1995). Aberrant driving behaviour: Errors and violations. *Ergonomics*, 38, 1759–1771.

Bollini, A. M., Walker, E. F., Hamann, S., and Kestler, L. (2004). The influence of perceived control and locus of control on the cortisol and subjective responses to stress. *Biological Psychology*, 67, 245–260.

Brown, I. D. (1962). Measuring the spare mental capacity of car drivers by a subsidiary auditory task. *Ergonomics*, 5, 247–250.

Brown, I. D., Tickner, A. H., and Simmonds, D. C. (1966). Effects of prolonged driving upon driving skill and performance of a subsidiary task. *Industrial Medicine and Surgery*, 35, 760–765.

Bursill, A. E. (1958). The restriction of peripheral vision during exposure to hot and humid conditions. *Quarterly Journal of Experimental Psychology*, 10, 114–129.

Cackowski, J. M., and Nasar, J. L. (2003). The restorative effects of roadside vegetation: Implications for automobile driver anger and frustration. *Environment and Behavior*, 35, 736–751.

Cartwright, S., Cooper, C. L., and Barron, A. (1996). The company car driver: Occupational stress as a predictor of motor vehicle accident involvement. *Human Relations*, 49, 195–208.

Cohen, S., Evans, G. W., Krantz, D. S., and Stokols, D. (1986). *Behavior, Health and Environmental Stress*. New York: Plenum.

Collantes, G. O., and Mokhtarian, P. L. (2007). Subjective assessments of personal mobility: What makes the difference between a little and a lot? *Transport Policy*, 14, 181–192.

Costa, G., Pickup, L., and Di Martino, V. (1988a). Commuting – a further stress factor for working people: Evidence from the European Community. *International Archives of Occupational and Environmental Health*, 60, 371–376.

(1988b). Commuting – a further stress factor for working people: Evidence from the European Community II. An empirical study. *International Archives of Occupational and Environmental Health*, 60, 377–385.

De Croon, E. M., Sluiter, J. K., Blonk, R. W. B., Broesen, J. P. J., and Frings-Dresen, M. H. W. (2004). Stressful work, psychological job strain, and turnover: A 2-year prospective cohort study of truck drivers. *Journal of Applied Psychology*, 89, 442–454.

Dubos, R. (1965). *Man Adapting*. New Haven, CT: Yale University Press.

(1969). The human environment. *Science Journal*, Oct., 75–80.

Ellaway, A., Macintyre, S., Hiscock, R., and Kearns, A. (2003). In the driving seat: Psychosocial benefits from private motor vehicle transport compared to public transport. *Transportation Research Part F*, 6, 217–231.

Ellison, N. B. (1999). Social impacts: New perspectives on telework. *Social Science Computer Review*, 17, 338–356.

European Commission. (2006). *Energy and Transport: Report 2000–2004*. Brussels: Directorate General for Energy and Transport.

Evans, G. W. (1994). Working on the hot seat: Urban bus operators. *Accident Analysis & Prevention*, 26, 181–193.

Evans, G. W., and Carrere, S. (1991). Traffic congestion, perceived control, and psychophysiological stress among urban bus drivers. *Journal of Applied Psychology*, 76, 658–663.

Evans, G. W., and Wener, R. E. (2006). Rail commuting duration and passenger stress. *Health Psychology*, 25, 408–412.

Evans, G. W., Wener, R. E., and Phillips, D. (2002). The morning rush hour: Predictability and commuter stress. *Environment and Behavior*, 34, 521–530.

Fink, G. (2000). *Encyclopedia of Stress. Volumes 1–3.* San Diego, CA: Academic Press.

Finkelman, J. M., and Glass, D. C. (1970). Reappraisal of the relationship between noise and human performance by means of a subsidiary task measure. *Journal of Applied Psychology*, 54, 211–213.

Finkelman, J. M., Zeitlin, L. R., Filippi, J. A., and Friend, M. A. (1977). Noise and driver performance. *Journal of Applied Psychology*, 62, 713–718.

Folkman, S. (1984). Personal control and stress and coping processes: A theoretical analysis. *Journal of Personality and Social Psychology*, 46, 839–852.

Frazier, P., Steward, J., and Mortensen, H. (2004). Perceived control and adjustment to trauma: A comparison across events. *Journal of Social and Clinical Psychology*, 23, 303–324.

Gardner, B., and Abraham, C. (2007). What drives car use? A grounded theory analysis of commuters' reasons for driving. *Transportation Research Part F*, 10, 187–200.

Gatersleben, B., and Uzzell, D. (2007). Affective appraisals of the daily commute: Comparing perceptions of drivers, cyclists, walkers, and users of public transport. *Environment and Behavior*, 39, 416–431.

Glass, D. C., and Singer, J. E. (1972). *Urban Stress: Experiments on noise and social stressors*. New York: Academic Press.

Golden, T. D. (2006). Avoiding depletion in virtual work: Telework and the intervening impact of work exhaustion on commitment and turnover intentions. *Journal of Vocational Behavior*, 69, 176–187.

Golden, T. D., Veiga, J. F., and Simsek, Z. (2006). Telecommuting's differential impact on work–family conflict: Is there no place like home? *Journal of Applied Psychology*, 91, 1340–1350.

Greenberger, D. B., Porter, G., Miceli, M. P., and Strasser, S. (1991). Responses to inadequate personal control in organizations. *Journal of Social Issues*, 47, 111–128.

Grieco, M., Pickup, L., and Whipp, R. (1989). *Gender, Transport, and Employment: The impact of travel constraints*. Aldershot, UK: Avebury.

Gulian, E., Glendon, A. I., Matthews, G., Davies, D. R., and Debney, L. M. (1990). The stress of driving: A diary study. *Work and Stress*, 4, 7–16.

Gulian, E., Matthews, G., Glendon, A. I., Davies, D. R., and Debney, L. M. (1989). Dimensions of driver stress. *Ergonomics*, 32, 585–602.

Handy, S., Weston, L., and Mokhtarian, P. L. (2005). Driving by choice or necessity? *Transportation Research Part A*, 39, 183–203.

Hartig, T., Kylin, C., and Johansson, G. (2007). The telework tradeoff: Stress mitigation vs. constrained restoration. *Applied Psychology*, 56, 231–253.

Hartley, L. R., and El Hassani, J. (1994). Stress, violations and accidents. *Applied Ergonomics*, 25, 221–230.

Hennessy, D. A., and Wiesenthal, D. L. (1997). The relationship between traffic congestion, driver stress and direct versus indirect coping behaviours. *Ergonomics*, 40, 348–361.

(1999). Traffic congestion, driver stress, and driver aggression. *Aggressive Behavior*, 25, 409–423.

Hennessy, D. A., Wiesenthal, D. L., and Kohn, P. M. (2000). The influence of traffic congestion, daily hassles, and trait stress susceptibility on state driver stress: An interactive perspective. *Journal of Applied Biobehavioral Research*, 5, 162–179.

Hentschel, U., Bijleveld, C., Kiessling, M., and Hosemann, A. (1993). Stress-related psychophysiological reactions of truck drivers in relation to anxiety, defense, and situational factors. *Accident Analysis & Prevention*, 25, 115–121.

Hill, E. J., Hawkins, A. J., and Miller, B. C. (1996). Work and family in the virtual office: Perceived influences of mobile telework. *Family Relations*, 45, 293–301.

Hill, J. D., and Boyle, L. N. (2007). Driver stress as influenced by driving maneuvers and roadway conditions. *Transportation Research Part F*, 10, 177–186.

Hiscock, R., Macintyre, S., Kearns, A., and Ellaway, A. (2002). Means of transport and ontological security: Do cars provide psycho-social benefits to their users? *Transportation Research Part D*, 7, 119–135.

Jain, J., and Lyons, G. (2008). The gift of travel time. *Journal of Transport Geography*, 16, 81–89.

Jones, G. W. (2002). Southeast Asian urbanization and the growth of mega-urban regions. *Journal of Population Research*, 19, 119–136.

Kluger, A. V. (1998). Commute variability and strain. *Journal of Organizational Behavior*, 19, 147–165.

Konradt, U., Hertel, G., and Schmook, R. (2003). Effects of management by objectives on perceived stress and job-satisfaction of teleworkers. *European Journal of Work and Organizational Psychology*, 12, 61–80.

Konradt, U., Schmook, R., and Mälecke, M. (2000). Impacts of telework on individuals, organizations and families – a critical review. *International Review of Industrial and Organizational Psychology*, 15, 99.

Kontogiannis, T. (2006). Patterns of driver stress and coping strategies in a Greek sample and their relationship to aberrant behaviors and traffic accidents. *Accident Analysis and Prevention*, 38, 913–924.

Koslowsky, M., Aizer, A., and Krausz, M. (1996). Stressor and personal variables in the commuting experience. *International Journal of Manpower*, 17, 4–14.

Koslowsky, M., Kluger, A. V., and Reich, M. (1995). *Commuting Stress: Causes, effects, and methods of coping*. New York: Plenum Press.

Koslowsky, M., and Krausz, M. (1993). On the relationship between commuting, stress symptoms, and attitudinal measures: A LISREL application. *Journal of Applied Behavioral Science*, 29, 485–492.

Kossek, E. E., Lautsch, B. A., and Eaton, S. C. (2006). Telecommuting, control, and boundary management: Correlates of policy use and practice, job control, and work–family effectiveness. *Journal of Vocational Behavior*, 68, 347–367.

Lagarde, E., Chastang, J. F., Gueguen, A., Coeuret-Pellicer, M., Chiron, M., and Lafont, S. (2004). Emotional stress and traffic accidents: The impact of separation and divorce. *Epidemiology*, 15, 762–766.

Lazarus, R. S. (1966). *Psychological Stress and the Coping Process*. New York: McGraw-Hill.

Lazarus, R. S., and Folkman, S. (1984). *Stress, Appraisal, and Coping.* New York: Springer.

Lefcourt, H. (1973). The function of the illusions of control and freedom. *American Psychologist*, 28, 417–425.

Legree, P. J., Heffner, T. S., Psotka, J., Martin, D. E., and Medsker, G. J. (2003). Traffic crash involvement: Experiential driving knowledge and stressful contextual antecedents. *Journal of Applied Psychology*, 88, 15–26.

Lewis, S., and Cooper, C. L. (1995). Balancing the work/home interface: A European perspective. *Human Resource Management Review*, 5, 289–305.

Lundberg, U., and Lindfors, P. (2002). Psychophysiological reactions to telework in female and male white-collar workers. *Journal of Occupational Health Psychology*, 7, 354–364.

Lyons, G., and Chatterjee, K. (2008). A human perspective on the daily commute: Costs, benefits and trade-offs. *Transport Reviews*, 28, 181–198.

Lyons, G., and Urry, J. (2005). Travel time use in the information age. *Transportation Research Part A*, 39, 257–276.

Macintyre, S., Hiscock, R., Kearns, A., and Ellaway, A. (2001). Housing tenure and car access: Further exploration of the nature of their relations with health in a UK setting. *British Medical Journal*, 55, 330.

Mackie, R. R., O'Hanlon, J., and McCauley, M. (1974). *A Study of Heat, Noise and Vibration in Relation to Driver Performance and Physiological Status.* DOT report no. HS-801 313. Washington, DC: US Department of Transportation.

Mann, E., and Abraham, C. (2006). The role of affect in UK commuters' travel mode choices: An interpretative phenomenological analysis. *British Journal of Psychology*, 97, 155–176.

Marsh, P., and Collett, P. (1986). *Driving Passion: The psychology of the car.* London: Jonathan Cape.

Matthews, G., Tsuda, A., Xin, G., and Ozeki, Y. (1999). Individual differences in driver stress vulnerability in a Japanese sample. *Ergonomics*, 42, 401–415.

McLennan, P., and Bennetts, M. (2003). The journey to work: A descriptive UK case study. *Facilities*, 21, 180–187.

Mokhtarian, P. L. (1990). A typology of relationships between telecommunications and transportation. *Transportation Research Part A*, 24, 231–242.

Mokhtarian, P. L., and Salomon, I. (2001). How derived is the demand for travel? Some conceptual and measurement considerations. *Transportation Research Part A*, 35, 695–719.

Mokhtarian, P. L., Salomon, I., and Redmond, L. S. (2001). Understanding the demand for travel: It's not purely "derived." *Innovation*, 14, 355–380.

Nilles, J. M. (1975). Telecommunications and organizational decentralization. *Communications, IEEE Transactions on Legacy, pre-1988*, 23, 1142–1147.

Novaco, R. W., and Collier, C. (1994). Commuting stress, ridesharing and gender: Analyses from the 1993 State of the Commute Study in southern California. *Transportation Research Record*, 1433, 170–176.

Novaco, R. W., Kliewer, W., and Broquet, A. (1991). Home environment consequences of commute travel impedance. *American Journal of Community Psychology*, 19, 881–909.

Novaco, R. W., Stokols, D., Campbell, J., and Stokols, J. (1979). Transportation, stress, and community psychology. *American Journal of Community Psychology*, 7, 361–380.

Novaco, R. W., Stokols, D., and Milanesi, L. (1990). Objective and subjective dimensions of travel impedance as determinants of commuting stress. *American Journal of Community Psychology*, 18, 231–257.

Ory, D. T., and Mokhtarian, P. L. (2005). When is getting there half the fun? Modeling the liking for travel. *Transportation Research Part A*, 39, 97–123.

Parker, D., Reason, J. T., Manstead, A. S. R., and Stradling, S. G. (1995). Driving errors, driving violations and accident involvement. *Ergonomics*, 38, 1036–1048.

Parsons, R., Tassinary, L. G., Ulrich, R. S., Hebl, M. R., and Grossman-Alexander, M. (1998). The view from the road: Implications for stress recovery and immunization. *Journal of Environmental Psychology*, 18, 113–140.

Pisarski, A. E. (1987). *Commuting in America: National report on commuting patterns and trends*. Westport, CT: ENO Foundation for Transportation.

(2006). *Commuting in America: The third national report on commuting patterns and trends*. Washington, DC: Transportation Research Board.

Plutchik, R. (1959). The effects of high intensity intermittent sound on performance, feeling, and physiology. *Psychological Bulletin*, 56, 133–151.

Raggatt, P. T. F. (1991). Work stress among long-distance coach drivers: A survey and correlational study. *Journal of Organizational Behavior*, 12, 565–579.

Reason, J., Manstead, A., Stradling, S., Baxter, J., and Campbell, K. (1990). Errors and violations on the roads: A real distinction. *Ergonomics*, 33, 1315–1332.

Redmond, L. S., and Mokhtarian, P. L. (2001). The positive utility of the commute: Modeling ideal commute time and relative desired commute amount. *Transportation*, 28, 179–205.

Rosenbloom, S. (1992). Why working families need a car. In M. Wachs and M. Crawford (Eds.), *The Car and the City: The automobile, the built environment, and daily urban life* (pp. 39–56). Ann Arbor: University of Michigan Press.

Rothengatter, T., and Vaya, E. C. (1997). *Traffic and Transport Psychology: Theory and application*. New York: Pergamon.

Schaeffer, M. H., Street, S. W., Singer, J. E., and Baum, A. (1988). Effects of control on the stress reactions of commuters. *Journal of Applied Psychology*, 18, 944–957.

Selye, H. (1956). *The Stress of Life*. New York: McGraw-Hill.

Shimomitsu, T. (1999). Occupational stress and health in Japan. Finnish Institute of Occupational Health. Retrieved July 2008, from www.ttl.fi/Internet/English/Information/Electronic+journals/Asian-Pacific+Newsletter/1999-01/08.htm.

Simon, F., and Corbett, C. (1996). Road traffic offending, stress, age, and accident history among male and female drivers. *Ergonomics*, 39, 757–780.

Singer, J. E., Lundberg, U., and Frankenhauser, M. (1977). Stress on the train: A study of urban commuting. In A. Baum, J. E. Singer, and S. Valins (Eds.), *Advances in Environmental Psychology* (pp. 41–56). New York: Erlbaum.

Sluiter, J. K., van Der Beek, A. J., and Frings-Dresen, M. H. W. (1998). Work stress and recovery measured by urinary catecholamines and cortisol excretion in long distance coach drivers. *Occupational and Environmental Medicine*, 55, 407–413.

Spielberger, C. D. (1996). *State-Trait Anger Expression Inventory: STAXI professional manual.* Tampa, FL: Psychological Assessment Resources.

Standen, P. (2000). The home/work interface. In K. Daniels, D. A. Lamond, and P. Standen (Eds.), *Managing Telework: Perspectives from human resource management and work psychology* (pp. 83–92). London: Business Press.

Standen, P., Daniels, K., and Lamond, D. (1999). The home as a workplace: Work–family interaction and psychological well-being in telework. *Journal of Occupational Health Psychology*, 4, 368–381.

Steg, L. (2005). Car use: Lust and must. Instrumental, symbolic and affective motives for car use. *Transportation Research Part A*, 39, 147–162.

Steg, L., Vlek, G., and Slotegraaf, G. (2001). Instrumental-reasoned and symbolic-affective motives for using a motor car. *Transportation Research Part F*, 4, 151–169.

Stephens, G. K., and Szajna, B. (1998). Perceptions and expectations: Why people choose a telecommuting work style. *International Journal of Electronic Commerce*, 3, 70–85.

Stokes, G., and Hallett, S. (1992). The role of advertising and the car. *Transport Reviews*, 12, 171–183.

Stokols, D., and Novaco, R. W. (1981). Transportation and well-being. In I. Altman, J. F. Wohlwill, and P. B. Everett (Eds.), *Transportation and Behavior* (pp. 85–130). New York: Plenum.

Stokols, D., Novaco, R. W., Stokols, J., and Campbell, J. (1978). Traffic congestion, type-A behavior, and stress. *Journal of Applied Psychology*, 63, 467–480.

Trent, J. T., Smith, A. L., and Wood, D. L. (1994). Telecommuting: Stress and social support. *Psychological Reports*, 74, 1312–1314.

Troup, C., and Dewe, P. (2002). Exploring the nature of control and its role in the appraisal of workplace stress. *Work & Stress*, 16, 335–355.

Veneziano, C., Veneziano, L., and Fichter, M. (1994). Stress-related factors associated with driving while intoxicated. *Journal of Alcohol and Drug Education*, 39, 87–98.

Vivoli, G., Bergomi, M., Rovesti, S., Carrozzi, G., and Vezzosi, A. (1993). Biochemical and haemodynamic indicators of stress in truck drivers. *Ergonomics*, 36, 1089–1097.

Wachs, M., and Crawford, M. (Eds.) (1992). *The Car and the City: The automobile, the built environment, and daily urban life*. Ann Arbor: University of Michigan Press.

Westerman, S. J., and Haigney, D. (2000). Individual differences in driver stress, error and violation. *Personality and Individual Differences*, 29, 981–998.

White, S. M., and Rotton, J. (1998). Type of commute, behavioral aftereffects, and cardiovascular activity: A field experiment. *Environment and Behavior*, 30, 763–780.

Wiesenthal, D. A., Hennessy, D. A., and Totten, B. (2000). The influence of music on driver stress. *Journal of Applied Social Psychology*, 30, 1709–1719.

WorldatWork. (2007). *Telework Trendlines for 2006*. Retrieved July 2008, from www.workingfromanywhere.org/news/Trendlines_2006.pdf.
Wyon, D. P., Wyon, I., and Norin, F. (1996). Effects of moderate heat stress on driver vigilance in a moving vehicle. *Ergonomics*, 39, 61–75.
(2007). Telework Trendlines for 2006: 2007 survey brief. Scottsdale, AZ: WorldatWork and The Dieringer Research Group. Retrieved March 17, 2009, from www.workingfromanywhere.org/news/Trendlines_2006.pdf.

8 Technology and medicine

Jeffrey W. Jutai, Sherry Coulson, and Elizabeth Russell-Minda

Medical technology in the broadest sense refers to the diagnostic or therapeutic application of science and technology to improve the management of health conditions, and describes a vast array of devices and techniques. Technologies may be behavioral, biological, educational, electronic, mechanical, or pharmacological, or other methods to increase life span and/or improve the quality of life. It may seem pointless to make a sweeping generalization about the impact of medical technology on well-being given this wide range of technologies and different purposes for which they were created. Nonetheless, as it is generally assumed that virtually every form of medical technology has a positive impact on quality of life and well-being, it is important to analyze this assumption critically.

This chapter begins by examining how the relationship between technology and health is conceptualized and therefore how effectively it might be researched. It then examines the research evidence for impact on quality of life and well-being of technologies to assist mobility and vision for individuals with disabilities. It concludes with recommendations on how assistive technologies might be better researched and developed to enhance well-being.

Relevant constructs

Well-being and related constructs of quality of life and psychological functioning are part of a constellation that makes up the human experience with technology and so may be inseparable. There is no universally accepted definition of *quality of life* (QoL) but some approaches to definition lend themselves to measuring the impact of technology better than others. In our research on the impact of assistive devices for individuals who have a physical or sensory disability, we have adopted the position advocated by Renwick *et al.* (1994). These authors argue that QoL refers to the degree to which a person enjoys the important possibilities of his/her life. This is a definition that speaks directly to the theme of this book because it compels us to focus on the most important goals for technology in supporting human existence. In defining *well-being*, we refer to the degree to which individuals have positive appraisals and feelings about their lives, considered altogether or in terms of particular domains, such as health and recreation (Fuhrer, 2000). All three principal features of *psychological functioning* (conation, affect, and cognition) may be affected by technology (Fuhrer, 2000). Recent studies particularly examine how devices make users feel more competent, confident, and motivated to exploit life's possibilities (Jutai, 1999; Jutai and Day, 2002).

Assumptions about these constructs and how they are affected by technology include that they (1) are complex and multidimensional; (2) are dynamic, changing over time and over a person's life; (3) arise from the individual's interaction with his/her environment; and (4) are experienced differently from person to person, but have the same components for everyone (Jutai, 1999). Approaches to outcome measurement should be faithful to these assumptions. The relationship between technology and health is difficult to measure but not impossible (Fuhrer *et al.*, 2003; Jutai *et al.*, 2005).

Conceptual framework

Ohnabe (2006) describes a useful conceptual framework for health technology that distinguishes the QoL impact of *medical* and *assistive* technologies. This framework is presented in Figure 8.1. Medical technologies are more narrowly defined, and are designed for assessment and intervention at the level of physical health and healing or, in the language of the International Classification of Functioning, Disability and Health (ICF), *body function and structures* (World Health Organization, 2001). Medical devices are applied in one or more of the following areas: *life support* (e.g., dialysis, mechanical ventilator); *substitution for body structure or function* (e.g., implanted electrical stimulator, prosthetic limb, cochlear implant); *medical treatment* (e.g., injection devices); and *measurement* (e.g., glucose monitor, fMRI). Ohnabe's framework makes it clear that without technology where it is needed to support or prolong life, individuals may still have QoL but their life span will be shorter. At the same time, we do not expect medical devices in the narrow sense to directly and appreciably improve QoL and well-being. These improvements are much more influenced by the opportunities to engage in life activities and participate in society. In fact, at times there can be opportunity costs associated with constraints imposed by life support technology (e.g., dialysis and mechanical ventilation) that may actually curtail these opportunities and thereby diminish QoL and well-being.

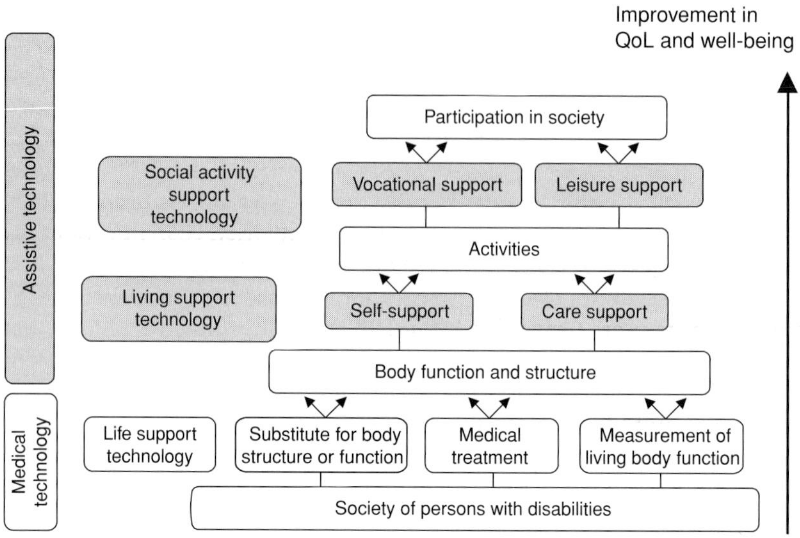

Figure 8.1 Conceptual framework for relating technology to improvement in quality of life and well-being.
Source: Adapted from Ohnabe, 2006.

Assistive technologies are designed for intervention at the component levels described in the ICF as *Activities and Participation*. There are devices to help individuals to perform activities through *self-support* (e.g., powered wheelchair, print magnifier, augmentative communication device) and *support from others* (e.g., toilet and bath transfer devices; assistive devices for feeding). Other technologies are designed to enhance participation in society through work (e.g., adapted computer workstation, head pointer, voice-operated equipment) and recreation (e.g., adapted sports equipment, closed captioned films and TV programs). In this framework, assistive technologies have more immediate consequences than medical technologies for the performance of life activities and participation in society and, hence, for improvement in QoL and well-being.

Assistive Technology Devices for mobility

Assistive Technology Devices (ATDs) are extremely prevalent forms of health care interventions for persons who have a physical disability. ATDs include both *assistive devices* (those applied to or directly manipulated by the person) and *special equipment* (attachments to the original structure of the physical environment) (Gitlin, 2002). An example of an assistive device would be a wheelchair, and an example of special equipment may be grab bars in the bathroom. An ATD is defined as "any item, piece of equipment, or product system, whether acquired commercially off the shelf, modified, or customized, that is used to increase, maintain, or improve functional capabilities of a [person] with a disability" (US Department of Education, 2004). There are currently more than 29,000 types of assistive devices that have been cataloged in the United States (Abledata, 2008). That number reaches infinity when customizations are factored in. The most important reasons for introducing an ATD into a person's life are to improve, or at least sustain, QoL, well-being, and dignity for its user (Gryfe and Jutai, 1998).

Categories of ATDs most commonly used by people with physical disabilities include: aids to assist activities of daily living; aids for adaptive and augmentative communication; computer access; electronic aids to daily living; hearing aids; home worksite modifications; orthotics and prosthetics; seating and positioning; vehicle modifications; vision aids; and wheelchair and mobility aids (Jutai, 2002). Physical disabilities requiring the use of an ATD for mobility cover a wide range of medical conditions, including, but not limited to: arthritis and orthopedic conditions, cerebral palsy, multiple sclerosis, amyotrophic lateral sclerosis, acquired brain injury, spinal cord injury, and stroke, as well as general mobility disabilities relating to aging.

Mobility is broadly defined as the ability to move from one place to another or from one position to another. Mobility limitations can impede the ability to perform activities of daily living requiring use of the lower limbs (Odding *et al.*, 2001). Generally speaking, mobility disabilities affect the function of the lower half of the body, but are often accompanied by limitations in other areas. ATDs used for mobility include canes (i.e., single-point, quad support, walking sticks), crutches (i.e., forearm or underarm crutches), walkers (i.e., rolling walkers, three-wheeled walkers), wheelchairs (i.e., manual, powered, tilt-in-space wheelchairs), scooters, and orthotics and prostheses (i.e., artificial legs, leg braces, foot arch supports). Also included under the auspice of ATDs used by people with mobility disabilities are a wide variety of support ATDs that assist in transferring from one place to another (i.e., hoists, lifts, shifting pads), personal support, hygiene, and safety (i.e., grab bars, commodes, seating cushions), as well as innumerable modifications to vehicles (i.e., hand gears), living spaces (i.e., modified houses including lowered countertops and wider doors), and the environment (i.e., ramps, accessibly designed public buildings). Mobility ATDs cover a wide range of technological complexity, from low-technology devices, such as canes, to high-technology devices such as powered wheelchairs.

ATDs, including aids to mobility such as canes, walkers, and wheelchairs, can be useful for accommodating the limitations in daily activities that result from a variety of disabling conditions. The Institute of Medicine (2007) reports that 40–50 million people in the US, or one in seven, have some form of disability. Over 15 million Americans with disabilities use medically related ATDs. Of those, approximately 7 million use ATDs for mobility (Carlson and Ehrlich, 2005).

Mobility ATDs have the potential to reduce residual disability, delay decline in function, decrease burden of care, and lower health care costs (Agree, 1999; Bateni and Maki, 2005; Gitlin *et al.*, 2001; Mann *et al.*, 1999; Verbrugge *et al.*, 1997). The importance and practical use of ATDs can be expected to increase over time as the global population ages and as public policy shifts to earlier community reintegration and places greater emphasis on self-care and care delivered in the home and community.

The impact of ATDs

Rehabilitation professionals and researchers assume that the introduction of an ATD should make a significant, measurable impact on the user's QoL and well-being above the effects of concurrent therapies, such as physical rehabilitation, occupational therapy, recreation therapy, surgeries, and medicines. These impacts are most often associated with improved functioning and greater independence; however, the impacts of ATDs also cover the many facets of psychological well-being. It is important how individuals, as well as the society around them, perceive their ATD. Often, these impacts have little to do with functional improvement. For example, a minor improvement in the cosmetic appearance of a walker may significantly improve user acceptance, leading to increased use and subjective well-being, without affecting an individual's mobility capacity.

Subjective well-being, as it relates to rehabilitation and assistive technology, refers to the degree to which an ATD affects how users appraise and feel about their lives, considered in terms of specific areas in their lives or overall (Fuhrer, 2000). It includes the extent to which ATDs and their outcomes are accepted by users and the effect on their lives. Satisfaction is a vital aspect of subjective well-being, and includes satisfaction with the device and services related to device use as well as the effects of the device on particular life domains and life in general. In the main, research has shown that ATDs positively enhance the QoL of people with disabilities (Davies *et al.*, 2003; Fuhrer *et al.*, 2003; Scherer, 1996). It should be noted though that the usage of assistive devices can sometimes give rise to negative feelings and be experienced as stigmatizing (Gitlin *et al.*, 1998; Wilson, 2005). This can probably be taken as an expression of a decreased QoL.

In medical research, QoL is couched in terms of it being "health related." *Health Related Quality of Life* (HRQoL) has been defined as "the value assigned to duration of life as modified by impairments, functional status, perceptions and opportunities influenced by disease, injury, treatment and policy" (Patrick and Erickson, 1993: 491). There is general agreement that HRQoL measures need to incorporate physical, mental (emotional and cognitive), social, and role functioning as well as an individual's perception of health and well-being (Berzon *et al.*, 1993; Bowling, 1995; Fitzpatrick *et al.*, 1992).

Traditional HRQoL measures, both generic and disease specific, could be used to assess the impact of an ATD on the lives of its users; however, these measures are limited as they are typically too medically oriented. Most ATDs are not intended to promote health and healing, but rather improve or restore functional capabilities. Existing measures are also designed to assess the health status or change in health status – not the impact attributable to any particular form of intervention. Furthermore, ATDs are meant to achieve different QoL and well-being outcomes from surgery, physical therapy, and pharmaceutical treatments (see Figure 8.1).

The use of ATDs, from the perspective of health care providers, is considered desirable because it is assumed that use reduces the risk for additional disability and illness, reduces demand for caregiver and community resources, and reduces costs. However, from the perspective of the user, ATDs are important if they improve function, participation, and well-being (Jutai *et al.*, 2005).

Although tools to measure functional rehabilitation outcomes have been developed, these tools do not generally consider ATDs when rating the individual's functional performance. For example, the Functional Independence Measure (FIM), while widely used in rehabilitation outcomes research, is not directly applicable to ATDs. In fact, if the person being assessed uses any form of ATD, they are penalized because ATD use is assumed to be a burden. Other traditional tools used in rehabilitation do not recognize the efficacy of ATDs at all.

Instruments that are designed specifically for measuring the effects of ATDs include the Quebec User Evaluation of Satisfaction with Assistive Technology (QUEST) (Demers *et al.*, 2000), the Assistive Technology Device Predisposition Assessment (ATD PA) (Scherer *et al.*, 2005), the Assistive Technology Outcome Measure (ATOM) (Dharne *et al.*, 2006), and the Psychosocial Impact of Assistive Devices Scale (PIADS) (Day and Jutai, 1996).

Health conditions and mobility ATDs

Exploring the impacts of stroke is of vital importance to rehabilitation and health care research as this disease is recognized as a leading health problem in our aging society and is a leading cause of disability. According to Mayo (1996), two-thirds of stroke survivors will have varying degrees of disability. Stroke has consequentially been classified as the most disabling chronic disease that affects not only the individual but also the society (Mayo *et al.*, 2002). While there is a large body of research exploring the recovery process poststroke, most focuses on consequences related to neurological impairments, functional disabilities, and loss of independence, and little has focused on QoL and the role of ATDs in the lives of those affected by stroke.

Stroke survivors are frequent users of ATDs which aim to accommodate disabilities that affect mobility, orientation, activities of daily living, and self-care. Older adults living at home after a stroke own an average of almost sixteen ATDs (Mann *et al.*, 1995). To date, little research has been conducted to support or refute the claim that assistive devices increase independence or improve the QoL of users (McWilliam *et al.*, 2000). Promoting independence is important to both stroke survivors and their caregivers. The use of assistive technology devices is one approach to enhancing independence.

A recent study of 316 Canadian stroke survivors (135 of whom received a mobility ATD poststroke, and 181 did not) found that the mobility devices prescribed after stroke were well matched to the functional status of the user. Canes, walkers, and wheelchairs were commonly procured devices, and canes were used as a primary aid to mobility more often than the other categories of device. Wheelchair users were more physically impaired than cane and walker users (Jutai *et al.*, 2007). Given that the device users in this study were early users, as defined by Gitlin *et al.* (1996), the relationships between device types and functional status are expected to change over time.

Multiple sclerosis (MS) is a chronic neurological disease that affects all aspects of personal functioning, including mobility. Mobility aids are the most commonly reported types of ATDs used by people with MS (Finlayson *et al.*, 2001). Little research to date has explored the use of ATDs by people with MS and the impact of these devices on the user. One study found that wheelchair use by people with MS was positively related to QoL (Ford *et al.*, 2001). Another study, using the PIADS to measure the psychosocial impact of wheelchair use for those with MS, interviewed sixteen hospitalized adults, whose ATD use covered a wide range of wheelchair types and patterns of device use. Findings suggested that the use of wheelchairs had a positive effect on those with MS (Devitt *et al.*, 2003).

Amyotrophic lateral sclerosis (ALS) is an adult onset, progressive neuromuscular disease that affects all aspects of daily living. Jutai and Gryfe (1998) found that wheelchairs were the most commonly prescribed ATDs for those with ALS. Patients with ALS also used other aids to assist with communication and activities of daily living. Using the PIADS, the authors found that the impact of all types of ATDs used was positive, as reported by both the users and their caregivers.

Aging and mobility ATDs

In the United States, it is estimated that two-thirds of older adults with disabilities use at least one ATD (Agree and Freedman, 2000). Wheelchairs are among the most prevalent forms of assistive technology used by older adults who have a physical disability (Jutai *et al.*, 2001; Wressle and Samuelsson, 2004). The use of wheelchairs and other ATDs for mobility by older adults results from numerous health conditions and general decline associated with the aging process. ATDs for mobility can enhance the psychological well-being and QoL of elderly individuals who use them. In the United States, a study of the prevalence of ATD use found that 14–18 percent of adults over the age of 65 used at least one ATD (Cornman *et al.*, 2005). A Dutch study of adults aged 70–89 found that 89 percent possessed at least one ATD and 80.6 percent used those devices (Roelands *et al.*, 2002). The mean number of devices owned was three, with canes being most prevalent, but interestingly, they were not the most used device. The nonuse of devices was positively related to various sociocognitive factors. In an American study, Gitlin *et al.* (1996) found that, of 642 older adult (>55) ATD users (considered early users) discharged from hospital after a stroke, amputation, or general orthopedic conditions, 50 percent used their prescribed ATDs all of the time. However, this study could not conclude what factors were related to abandonment of a prescribed device.

Mobility ATDs make up a huge portion of health and rehabilitation costs; however, there is a high percentage of abandonment of these devices, suggesting that there is a failure to fit the device with the individual (Day *et al.*, 2002). A number of factors contribute to the nonuse of an ATD, including: personal factors (i.e., demographic characteristics, level of disability); device-specific factors (i.e., quality, appearance); environmental factors (i.e., social support, barriers, opportunities); and intervention-related factors (i.e., instruction and training, correct prescription, follow-up). The need for an ATD alone does not indicate its use, and, in fact, the influence of personal meanings attached to ATDs on the integration of ATDs into a person's life plays an important role in ATD use (Pape *et al.*, 2002). In a study of 227 adults with disabilities, Phillips and Zhao (1993) found that mobility ATDs were abandoned more frequently than other types of devices. Four factors leading to nonuse were identified: involvement in the selection of the device; ease of getting a device; poor device performance; and a change in user need or performance (Phillips and Zhao, 1993). In a Swedish study of satisfaction with wheelchairs and walkers, Wressle and Samuelsson (2004) found that, while user satisfaction was generally high, the primary area of dissatisfaction was service follow-up. Understanding the role of satisfaction (and dissatisfaction), both with devices themselves and with related services, is important in understanding the continued use or abandonment of a device (Scherer *et al.*, 2005).

None of the studies reviewed used experimental, randomized controlled studies to examine whether the effects on QoL, satisfaction, and well-being could be attributed specifically to the use of ATDs. As the global population ages, the use of ATDs and their importance for their users and society will increase significantly and therefore research into assistive technology and QoL will become increasingly important. If applied effectively, ATDs can be a key approach to promoting the functional recovery and long-term health of people with mobility disabilities, as well as improving the QoL of those users.

Assistive technology for low vision and blindness

Recent global estimates from the World Health Organization (WHO) indicate that approximately 161 million people have some form of visual impairment. Of these individuals 124 million are categorized as having low vision, and 37 million are blind (Dandona and Dandona, 2006). This estimate is based on definitions outlined in the tenth revision of the WHO *International Statistical Classification of Diseases* (2003), in which low vision is defined as visual acuity of less than 6/18, but equal or better than 3/60, or corresponding visual field loss to less than 20 degrees remaining, in the better eye with best possible correction. Blindness is defined as visual acuity of less than 3/60, or corresponding visual field loss to less than 10 degrees remaining, in the better eye with best possible correction. In order to maintain independence with the activities of daily life and continue to lead productive and fulfilling lives, many individuals with visual impairment seek vision rehabilitation services, which typically incorporate training in the use of low-vision aids as well as other types of educational strategies. The most common type of assistive technology for vision rehabilitation includes optical aids for reading and other tasks, such as magnifiers, electronic magnification systems, and Closed Circuit Televisions (CCTVs). Prisms may help with field enhancement and telescopes may be useful for distance viewing and spotting tasks. Implementing proper lighting conditions at home or work are simple ways to modify one's environment. Sight substitution technologies include the use of Braille and other tactile materials. Assistive technology can also include orientation and mobility training (including device training), and driver rehabilitation. Ultimately, the desired outcomes of receiving vision rehabilitation are for individuals to attain the maximum function of any remaining vision, increase levels of functional ability, increase independence, and improve their overall QoL (AHRQ, 2004; Jutai *et al.*, 2005).

Assistive technology for vision loss has typically been grouped by functional application, by specific visual impairment condition, or by device type. Some devices are multifunctional – so possibly more resistant to simple categorization – and therefore a taxonomy for low-vision assistive technology might be proposed to organize these devices and strategies according to the way in which they either "enhance" or "substitute" vision. *Sight enhancement* assistive technology can be defined as devices or strategies used to take advantage of useful residual vision. *Sight substitution* assistive technology may be designed around the premise that no useful or contributing vision is available, where individuals rely on non-vision senses of hearing and touch. Often, the success or effectiveness of the assistive technology is determined by how well it performs for the user and the level of satisfaction the individual has with the device (Jutai *et al.*, 2005, 2008). Outcome measures for determining the effectiveness of assistive devices or strategies can include both subjective and objective measures of performance. Subjective outcome measures – such as preference for a particular device, ease of use, or satisfaction with a certain type of technology – may reliably assess the impact a device may have on an individual's QoL and happiness with assistive technology. Subjective measures are typically assessed using psychometrically validated questionnaires or scales, which are administered by a clinician or therapist, or completed by the patients themselves. In addition, research syntheses using systematic review methods have helped to determine the strength of evidence for the effectiveness of assistive technology in improving the QoL and well-being of individuals with vision loss (Hooper *et al.*, 2008; Jutai *et al.*, 2008; Virgili and Acosta, 2006; Virgili and Rubin, 2006).

Rehabilitation research has assessed the impact of vision rehabilitation programs on improving individual QoL and well-being, activities of daily living, and the use of low-vision aids (Eklund *et al.*, 2004; McCabe *et al.*, 2000; Pankow *et al.*, 2004; Scanlan and Cuddeford, 2004). Brody *et al.* (1999) conducted a randomized clinical trial for evaluating the effects of a self-management program for individuals with age-related macular degeneration, on measures of self-efficacy, psychological stress, and use of low-vision aids. Mean QoL ratings were taken from Quality of Well-Being scale scores. The results of the short, weekly, two-hour group sessions showed substantially reduced psychological stress and improved self-efficacy for those who participated in the intervention. The self-management intervention also had a small effect on the increased use of low-vision aids in sixty-eight subjects. In another randomized controlled trial, Reeves *et al.* (2004) compared the effectiveness of an enhanced low-vision rehabilitation program with a conventional type of program, incorporating the training and use of low-vision aids. The enhanced program included additional home visits by a rehabilitation specialist who provided assistance with lighting modifications and guidance on suitable low-vision aids. The main outcome measures were vision-specific QoL, measured by the "vision core module 1" (VCM1) and generic health-related QoL (SF-36), psychological adjustment to vision loss, measured task performance, and restriction in everyday activities. The study results found no evidence supporting the benefits of an enhanced program over a conventional type of program via measures of subjective QoL ratings; however, the use of low-vision aids was high throughout the trial and increased with duration of follow-up.

Prisms and field enhancement devices can be beneficial prescriptions for people with visual field loss resulting from hemianopsia and unilateral visual neglect, typically occurring after a stroke. Prismatic lenses have also been advocated for some patients with age-related macular degeneration, where fixation is shifted to the nondiseased areas of the retina. Fresnel prisms and mirrors may be used to provide enhanced awareness of obstacles in the affected visual field, and may be useful in rehabilitation. Smith *et al.* (2005) examined the relative effectiveness of prism spectacles for subjects with age-related macular degeneration versus a control with no prism. Subjective outcome measures included assessments of QoL using the twenty-five-item National Eye Institute's Visual Function Questionnaire (NEI-VFQ-25), and activities of daily living with the Melbourne Low Vision Activities of Daily Living Index (MLVAI). The results indicated that the prism spectacles were no more effective than conventional glasses for improving reading performance, improving QoL, or increasing activities of daily living skills for individuals with age-related macular degeneration.

Monocular low-vision telescopes are commonly prescribed for individuals who require magnification to visualize distant objects of interest.

Most low-vision telescopes are either spectacle-mounted or handheld, and can be used for spotting, inspection, and scanning tasks. Bioptic telescopes are spectacle-mounted telescopic devices (monocular or binocular) that magnify objects located in the distance and can be used during driving for distance spotting tasks. Many jurisdictions and regions globally prohibit the use of bioptic telescope systems used specifically for driving, based on concern about the device impinging on the user's field of vision. Despite the controversies and conflicting views surrounding the use of bioptics for driving, some individuals note improvements in their overall feelings of well-being and QoL after training with the device because it has allowed them to continue to carry out their daily activities independently (Bowers *et al.*, 2005; Szlyk *et al.*, 2000).

Conventional optical low-vision aids, such as handheld or stand magnifiers, are popular, portable, multiapplication devices that can be used for a wide range of seeing tasks. One significant disadvantage is that higher-powered optical magnifiers necessitate very close working distances with highly constrained fields of view and significant optical aberrations (Culham *et al.*, 2004). When conventional low-vision aids produce insufficient magnification or cannot provide sustained performance for particular visual tasks, electronic magnification systems and CCTVs can provide high magnification with a good field of view. The eye-to-device working distance is larger than that of an optical magnifier, which allows for binocular viewing from a comfortable posture. With CCTVs and head-mounted or handheld video display systems (electronic vision enhancement systems), image size, brightness, and contrast can all be adjusted according to individual needs. Goodrich and Kirby (2001) compared patients' reading performance and preferences for their prescribed optical devices (stand magnifiers and microscopic lenses) with stand-mounted or handheld CCTVs. The results indicated significantly greater reading performance with CCTVs than with the patients' prescribed low-vision aids, and an overall preference for the stand-mounted CCTV over the handheld system. Similarly, Strong *et al.* (2003) noted that users of CCTVs had immediate psychosocial benefits after first receiving the device.

Peterson *et al.* (2003) examined whether reading and identification tasks (near vision) could be improved with various electronic vision enhancement systems, compared to the participants' own optical magnifiers. Although these systems provided benefits for users in terms of reading speed and acuity, some of the near-vision tasks could be performed just as fast using standard optical magnification. The results of a device-ranking questionnaire indicated that subjects found the *stand* electronic vision enhancement system, with monitor, easier to use than the *mouse* electronic vision enhancement system with monitor. They rated the mouse electronic vision enhancement system with the head-mounted device as having a difficulty similar to using their own magnifiers.

Among individuals with visual impairment, increased illumination has been shown to improve overall QoL and the ability to carry out daily activities. High-quality illumination includes not only a sufficient quantity of light but also conditions such as the direction of the light, the type of illumination, and the presence of glare. Several studies have shown that the reading performance of persons with age-related macular degeneration is more likely to improve with increased task illumination (Bowers *et al.*, 2001; Bullimore and Bailey, 1995; Eldred, 1992; Fosse and Valberg, 2004; LaGrow, 1986; Lovie-Kitchin *et al.*, 1983; Sloan, Habel, and Feiock, 1973). Optimal illumination for people with low vision can be measured using both objective evaluations (reading performance) and subjective preference methods (comfort ratings). Brunnström *et al.* (2004) evaluated the effects of suitable household lighting on subjective QoL, and the ability to carry out activities of daily living by increasing the lighting in various rooms of the home. They found a marked improvement in QoL and activities of daily living performance as a result of the lighting adjustments (as measured with a quality of well-being scale). Bowers *et al.* (2001) assessed individuals with age-related macular degeneration on reading performance without low-vision devices, for a range of print sizes at six different illuminance levels: 50, 300, 600, 1,000, 2,000, and 5,000 lux (a typical living-room illuminance being 50 lux). This study evaluated both objective optimal illuminance levels and subjective preferred illuminance levels. The majority of subjects preferred an illuminance level of 2,450 lux (median). The majority required task illumination of at least 2,000 lux to maximize reading performance, based on objective assessments.

Orientation and mobility research frequently uses measurements of mobility and obstacle detection performance, and typically takes place within laboratory-based settings (indoor or outdoor), or in "real-world" settings. Research related to subjective preferences and QoL assessments has typically been conducted with mobility aids (canes), rather than with specific types of training programs. In a recent systematic review of orientation and mobility training programs (Virgili and Rubin, 2006), the authors concluded that there is little evidence as to which type of orientation and mobility training is most beneficial for people with low vision who have specific needs and characteristics. In addition to the need for consensus on the adoption of standard measurement instruments which measure mobility performance, questionnaires which explore people's subjective experiences are also in need of development.

Traditional long straight canes are the most widely used mobility aid for individuals with visual impairment, and continue to be an effective mobility device (Rodgers and Emerson, 2005). Current research has not determined which canes produce the best performance, but users often express strong subjective preferences for particular features of canes, dependent upon environmental conditions (Ambrose-Zaken, 2005).

A number of adaptive technologies exist for both experienced Braille users as well as non-Braille users. Some of these technologies include refreshable Braille displays for reading or writing, Braille translation software, Braille embossers for producing hardcopy documents, portable note-taking devices, optical character recognition systems, screen magnification software, and speech output software. Sensory substitution, a concept introduced at the beginning of the 1970s, involves devices that acquire and process sensory information relating to the impaired system into signals that are adapted to a nonimpaired substitutive sensory system (Veraart *et al.*, 2004). Advances in tactile technology have also made maps more accessible for individuals with visual impairment by using tactile map materials (e.g., special inks and surfaces) (Siekierska *et al.*, 2003). For individuals with vision impairment, aspects of their QoL can be greatly improved with this type of assistive technology, but its usefulness is frequently dependent on their level of technical skill and access to standard or emerging technology. There is a paucity of research on these types of sensory and adaptive technology and their effects on psychological well-being and QoL.

Vision-specific instruments designed to measure patient satisfaction with devices as well as measures of QoL have been proved useful in providing valuable information to the low-vision research community. There is a strong case for using a wide range of outcome measures to describe the effectiveness of low-vision rehabilitation and assistive technology for individuals with visual impairments, at least until research studies identify which outcomes are the most important and nonredundant (Harper *et al.*, 1999). The use of these measurements can assist users, providers, and the research community with data on the practical utility and retention of low-vision assistive technology, based on psychosocial aspects. Subjective data can also help plan out proper economic analyses and provide the user with the tools for being informed consumers of devices, both in terms of practical costs and cosmesis. Outcome measures for evaluating subjective responses to low-vision assistive technology have been criticized for their lack of standardization or applicability. This may simply reflect the diversity of device types and ocular conditions, as well as variabilities among individuals and research environments.

Conclusions and recommendations

This chapter located studies related to medical technology, quality of life, and well-being, which incorporated a variety of interventions, comparisons, and outcomes. These different study methodologies can make it difficult to analyze or predict which strategy is the most effective for improving the well-being of individuals who have a functional limitation associated with mobility or vision. In addition, the studies selected for this review frequently evaluated subjects with multiple levels of impairment. This variability within the study populations makes it difficult to assess the utility of specific instruments for measuring QoL and well-being since, in many cases, particular intervention methods work better than others when levels of disability increase.

Various types of assistive devices are available to help people with their mobility and vision problems. None of these emerges as the most effective device because no device provides an identical level of functionality to another, nor do device users have precisely the same needs and expectations. Other differentiating factors include cost, ease of use, versatility, safety, universality, cosmetics, availability, serviceability, practicality, and adaptability.

It is evident from the results of this review that more randomized controlled trials and controlled studies with similar interventions and outcomes are needed. The ability to compare similar types of interventions and outcome measures within controlled study designs will allow researchers to conduct better evidence-based research syntheses, move recommendations into clinical practice, and potentially fill existing gaps in research. In addition, these findings are also important considerations when deciding on a particular type of assistive technology, especially for individuals with limited financial means.

Do people who use medical technologies have better well-being as a result of using these devices? There is no question that individuals can have life supported or prolonged by *medical* devices, but that does not necessarily translate to better well-being. There is also no doubt that, without their *assistive* devices, many individuals with disabilities and chronic conditions would have severe restrictions on their ability to perform life activities and participate in society, making it less likely that they would report having healthy QoL and well-being. The research evidence basis for this relationship though is generally weak but there are significant developments which should be encouraged. They include (1) rigorous research syntheses that apply common standards of evidence for well-being-related outcomes across the widest possible ranges of technologies (Hooper *et al.*, 2008; Jutai *et al.*, 2008); (2) improved outcome measures that capture meaningful information about the QoL and well-being of technology users (Lenker *et al.*, 2005); (3) improved designs for outcomes research that accommodate the dynamic and multivariate realities about the relationship between technology and well-being (Fuhrer *et al.*, 2003); and (4) improved conceptual frameworks for researching technology outcomes (Jutai *et al.*, 2005; Ohnabe, 2006).

References

Abledata. (2008). Web site. Retrieved January 28, 2008, from www.Abledata.com.

Agency for Healthcare Research and Quality (AHRQ) Technology Assessment Program. (2004). *Vision Rehabilitation for Elderly Individuals with Low Vision or Blindness*. Rockville, MD: Agency for Healthcare Research and Quality Technology Assessment Program. Retrieved March 2, 2009, from www.cms.hhs.gov/InfoExchange/Downloads/RTCvisionrehab.pdf.

Agree, E. (1999). The influence of personal care and assistive devices on the measurement of disability. *Social Science & Medicine*, 48, 427–443.

Agree, E. M., and Freedman, V. A. (2000). Incorporating assistive devices into community-based long-term care: An analysis of the potential for substitution and implementation. *Journal of Aging and Health*, 12, 426–450.

Ambrose-Zaken, G. (2005). Knowledge of and preferences for long cane components: A qualitative and quantitative study. *Journal of Visual Impairment & Blindness*, 99, 633–645.

Bateni, H., and Maki, B. E. (2005). Assistive devices for balance and mobility: Benefits, demands and adverse consequences. *Archives of Physical Medicine and Rehabilitation*, 86, 134–145.

Berzon, R., Hays, R. D., and Shumaker, S. A. (1993). International use, application and performance of health-related quality of life instruments. *Quality of Life Research*, 2, 367–368.

Bowers, A. R., Apfelbaum, D. H., and Peli, E. (2005). Bioptic telescopes meet the needs of drivers with moderate visual acuity loss. *Investigative Ophthalmology & Visual Science*, 46, 66–74.

Bowers, A. R., Meek, C., and Stewart, N. (2001). Illumination and reading performance in age-related macular degeneration. *Clinical and Experimental Optometry*, 84, 139–147.

Bowling, A. (1995). *Measuring Disease: A review of disease-specific quality of life measurement scales*. Buckingham: Open University Press.

Brody, B. L., Williams, R. A., Thomas, R. G., Kaplan, R. M., Chu, R. M., and Brown, S. I. (1999). Age-related macular degeneration: A randomized clinical trial of a self-management intervention. *Annals of Behavioral Medicine*, 21, 322–329.

Brunnström, G., Sörensen, S., Alsterstad, K., and Sjostrand, J. (2004). Quality of light and quality of life – the effect of lighting adaptation among people with low vision. *Ophthalmic and Physiological Optics*, 24, 274–280.

Bullimore, M. A., and Bailey, I. L. (1995). Reading and eye movements in age-related maculopathy. *Optometry and Vision Science*, 72, 125–138.

Carlson, D., and Ehrlich, N. (2005). *Assistive Technology and Information Technology Use and Need by Persons with Disabilities in the United States, 2001*. Washington, DC: National Institute on Disability and Rehabilitation Research, US Department of Education. Retrieved March 2, 2009, from www.ed.gov/rschstat/research/pubs/at-use/at-use-2001.pdf.

Cornman, J. C., Freedman, V. A., and Agree, E. M. (2005). Measurement of assistive device use: Implications for estimates of device use and disability in late life. *The Gerontologist*, 45, 347–358.

Culham, L., Chabra, A., and Rubin, G. (2004). Clinical performance of electronic, head-mounted, low-vision devices. *Ophthalmic and Physiological Optics*, 24, 281–290.

Dandona, L., and Dandona, R. (2006). What is the global burden of visual impairment? *BMC Medicine*, 4, 1–10.

Davies, A., DeSouza, L. H., and Frank, A. O. (2003). Changes in the quality of life of severely disabled people following provision of powered indoor/outdoor chairs. *Disability & Rehabilitation*, 25, 286–290.

Day, H., and Jutai, J. (1996). Measuring the psychosocial impact of assistive devices: The PIADS. *Canadian Journal of Rehabilitation*, 9, 159–168.

Day, H., Jutai, J., and Campbell, K. A. (2002). Development of a scale to measure the psychosocial impact of assistive devices: Lessons learned and the road ahead. *Disability & Rehabilitation*, 24, 31–37.

Demers, L., Weiss-Lambrou, R., and Ska, B. (2000). *Quebec User Evaluation of Satisfaction with Assistive Technology (Quest 2.0): An outcome measure for assistive technology devices*. Webster, NY: Institute for Matching Person and Technology.

Devitt, R., Chau, B., and Jutai, J. W. (2003). The effect of wheelchair use on the quality of life of persons with Multiple Sclerosis. *Occupational Therapy Practice and Research*, 17, 63–79.

Dharne, M., Lenker, J., Harris, F., and Sprigle, S. (2006). *Content Validity of Assistive Technology Outcome Measure (ATOM), Version 2.0*. Unpublished manuscript.

Eklund, K., Sonn, U., and Dahlin-Ivanoff, S. (2004). Long-term evaluation of a health education programme for elderly persons with visual impairment: A randomized study. *Disability & Rehabilitation*, 26, 401–409.

Eldred, K. B. (1992). Optimal illumination for reading in patients with age-related maculopathy. *Optometry and Vision Science*, 69, 46–50.

Finlayson, M., Guglielmello, L., and Liefer, K. (2001). Describing and predicting the possession of assistive devices among persons with multiple sclerosis. *The American Journal of Occupational Therapy*, 55, 545–551.

Fitzpatrick, R., Fletcher, A., Gore, S., Jones, D., Spiegelhalter, D., and Cox, D. (1992). Quality of life measures in health care. I: Applications and issues in assessment. *BMJ*, 305, 1074–1077.

Ford, H. L., Gerry, E., Johnson, M. H., and Tennant, A. (2001). Health status and quality of life of people with multiple sclerosis. *Disability & Rehabilitation*, 23, 516–521.

Fosse, P., and Valberg, A. (2004). Lighting needs and lighting comfort during reading with age-related macular degeneration. *Journal of Visual Impairment & Blindness*, 98, 389–409.

Fuhrer, M. J. (2000). Subjectifying quality of life as a medical rehabilitation outcome. *Disability & Rehabilitation*, 22, 481–489.

Fuhrer, M., Jutai, J. W., Scherer, M. J., and DeRuyter, F. (2003). A framework for the conceptual modeling of assistive technology outcomes. *Disability & Rehabilitation*, 25, 1243–1251.

Gitlin, L. N. (2002). Assistive technology in the home and community for older people: Psychological and social considerations. In M. J. Scherer (Ed.), *Assistive Technology: Matching device and consumer for successful rehabilitation* (pp. 109–122). Washington, DC: American Psychological Association.

Gitlin, L. N., Luborsky, M. R., and Schemm, R. L. (1998). Emerging concerns of older stroke patients about assistive device use. *Gerontologist*, 38(2), 169–180.

Gitlin, L. N., Mann, W., Tomita, M., and Marcus, S. M. (2001). Factors associated with home environmental problems among community living older people. *Disability & Rehabilitation*, 23, 777–787.

Gitlin, L. N., Schemm, R. L., Landsberg, L., and Burgh, D. (1996). Factors predicting assistive device use in the home by older people following rehabilitation. *Journal of Aging and Health*, 8, 554–575.

Goodrich, G. L., and Kirby, J. (2001). A comparison of patient reading performance and preference: Optical devices, handheld CCTV (Innoventions Magni-Cam), or stand-mounted CCTV (Optelec Clearview or TSI Genie). *Optometry*, 72, 519–528.

Gryfe, P., and Jutai, J. (1998). Assistive technologies: Clients' perceptions of impact on quality of life. *Rehabilitation and Community Care Management*, 7, 26–30.

Harper, R., Doorduyn, K., Reeves, B., and Slater, L. (1999). Evaluating the outcomes of low vision rehabilitation. *Ophthalmic Physiological Optics*, 19, 3–11.

Hooper, P., Jutai, J. W., Strong, G., and Russell-Minda, E. (2008). Age-related macular degeneration and low vision rehabilitation: A systematic review. *Canadian Journal of Ophthalmology*, 43(2), 180–187.

Institute of Medicine (IOM). (2007). *The Future of Disability in America*. Washington, DC: The National Academies Press.

Jutai, J. (1999). Quality of life impact of assistive technology. *Rehabilitation Engineering*, 14, 2–7.

(2002). Occupational therapy and assistive technology: The research challenge. *The Israel Journal of Occupational Therapy*, 11, E3–E22.

Jutai, J., Coulson, S., Teasell, R., Bayley, M., Garland, J., Mayo, N., and Wood-Dauphinee, S. (2007). Mobility assistive device utilization in a prospective study of patients with first-ever stroke. *Archives of Physical Medicine and Rehabilitation*, 88, 1268–1275.

Jutai, J., and Day, H. (2002). Psychosocial Impact of Assistive Devices Scale (PIADS). *Technology and Disability*, 14, 107–111.

Jutai, J., Day, H., Woolrich, W., and Strong, G. (2003). The predictability of retention and discontinuation of contact lenses. *Optometry*, 74, 299–307.

Jutai, J., Fuhrer, M. J., Demers, L., Scherer, M. J., and DeRuyter, F. (2005). Toward a taxonomy of assistive technology device outcomes. *American Journal of Physical Medicine & Rehabilitation*, 84, 294–302.

Jutai, J., and Gryfe, P. (1998). Impacts of assistive technology on clients with ALS. *Proceedings from the RESNA '98 Annual Conference* (pp. 54–65). Arlington, VA: RESNA Press.

Jutai, J., Hartley, M., Gryfe, P., and Day, H. (2001). Psychosocial impact of wheelchairs on elderly individuals who have a physical disability. Paper presented at International Conference on Technology and Aging, Toronto, September 12–14.

Jutai, J., Strong, G., and Russell-Minda, E. (2008). *Vision Rehabilitation Evidence-based Review: Assistive technologies for low vision and blindness*. The University of Western Ontario and CNIB. Retrieved March 2, 2009, from www.piads.ca/112/vrebr.htm.

LaGrow, S. J. (1986). Assessing optimal illumination for visual response accuracy in visually impaired adults. *Journal of Visual Impairment & Blindness*, 80, 888–895.

Lenker, J. A., Scherer, M. J., Fuhrer, M. J., Jutai, J. W., and DeRuyter, F. (2005). Psychometric and administrative properties of measures used in assistive technology device outcomes research. *Assistive Technology*, 17, 7–22.

Lovie-Kitchin, J. E., Bowman, K. J., and Farmer, E. J. (1983). Technical note: Domestic lighting requirements for elderly patients. *Australian Journal of Optometry*, 66, 93–97.

Mann, W. C., Hurren, D., Tomita, M., and Charvat, B. (1995). Assistive devices for home-based older stroke survivors. *Topics in Geriatric Rehabilitation*, 10, 75–76.

Mann, W. C., Ottenbacher, K. J., Fraas, L., Tomita, M., and Granger, C. V. (1999). Effectiveness of assistive technology and environmental interventions in maintaining independence and reducing home care costs for frail elderly: A randomized controlled trial. *Archives of Family Medicine*, 8, 210–217.

Mayo, N. E. (1996). Hospitalization and case-fatality rates for stroke in Canada from 1982 through 1991. The Canadian collaborative study group of stroke hospitalization. *Stroke*, 27, 1215–1220.

Mayo, N. E., Wood-Dauphinee, S., Cote, R., Durcan, L., and Carlton, J. (2002). Activity, participation, and quality of life 6 months poststroke. *Archives of Physical Medicine and Rehabilitation*, 83, 1035–1042.

McCabe, P., Nason, F., Demers, T. P., Friedman, D., and Seddon, J. M. (2000). Evaluating the effectiveness of a vision rehabilitation intervention using an objective and subjective measure of functional performance. *Ophthalmic Epidemiology*, 7, 259–270.

McWilliam, C. L., Diehl-Jones, W. L., Jutai, J. W., and Tadrissi, S. (2000). Care delivery approaches and seniors' independence. *Canadian Journal on Aging*, 19, 101–124.

Odding, E., Valkenburg, H. A., Stam, H. J., and Hofman, A. (2001). Determinants of locomotor disability in people aged 55 years and over: The Rotterdam study. *European Journal of Epidemiology*, 17, 1033–1041.

Ohnabe, H. (2006). Current trends in rehabilitation engineering in Japan. *Assistive Technology*, 18, 220–232.

Pankow, L., Luchins, D., Studebaker, J., and Chettleburgh, D. (2004). Evaluation of a vision rehabilitation program for older adults with visual impairment. *Topics in Geriatric Rehabilitation*, 20, 223–232.

Pape, L.-B. T., Kim, J., and Weiner, B. (2002). The shaping of individual meanings assigned to assistive technology: A review of personal factors. *Disability & Rehabilitation*, 24(1/2/3), 5–20.

Patrick, D. L., and Erickson, P. (1993). *Health Status and Health Policy*. New York: Oxford University Press.

Peterson, R. C., Wolffsohn, J. S., Rubinstein, M., and Lowe, J. (2003). Benefits of electronic vision enhancement systems (EVES) for the visually impaired. *American Journal of Ophthalmology*, 136, 1129–1135.

Phillips, B., and Zhao, H. (1993). Predictors of assistive technology abandonment. *Assistive Technology*, 5, 36–45.

Reeves, B. C., Harper, R. A., and Russell, W. B. (2004). Enhanced low vision rehabilitation for people with age related macular degeneration: A randomised controlled trial. *British Journal of Ophthalmology*, 88, 1443–1449.

Renwick, R., Brown, I., and Raphael, D. (1994). Quality of life: Linking a conceptual approach to service provision. *Journal of Developmental Disabilities*, 3, 32–44.

Rodgers, M. D., and Emerson, R. W. (2005). Human factor analysis of long cane design: Weight and length. *Journal of Visual Impairment & Blindness*, 99, 622–632.

Roelands, M., Van Oost, P., Depoorter, A. M., and Buysse, A. (2002). A social-cognitive model to predict the use of assistive devices for mobility and self-care in elderly people. *The Gerontologist*, 42, 39–54.

Scanlan, J. M., and Cuddeford, J. E. (2004). Low vision rehabilitation: A comparison of traditional and extended teaching programs. *Journal of Visual Impairment & Blindness*, 98, 601–611.

Scherer, M. J. (1996). Outcomes of assistive technology use on quality of life. *Disability & Rehabilitation*, 18, 439–448.

Scherer, M. J., Sax, C., Vanbiervliet, A., Cuschman, L. A., and Scherer, J. V. (2005). Predictors of assistive technology use: The importance of personal and psychosocial factors. *Disability & Rehabilitation*, 27, 1321–1331.

Siekierska, E., LaBelle, R., Brunet, L., McCurdy, B., Pulsifer, P., Rieger, M. K., and O'Neil, L. (2003). Enhancing spatial learning and mobility training of visually impaired people – a technical paper on the Internet-based tactile and audio-tactile mapping. *The Canadian Geographer*, 47, 480–493.

Sloan, L. L., Habel, A., and Feiock, K. (1973). High illumination as an auxiliary reading aid in diseases of the macula. *American Journal of Ophthalmology*, 76, 745–757.

Smith, H. J., Dickinson, C. M., Cacho, I., Reeves, B. C., and Harper, R. A. (2005). A randomized controlled trial to determine the effectiveness of prism spectacles for patients with age-related macular degeneration. *Archives of Ophthalmology*, 123, 1042–1050.

Strong, G., Jutai, J. W., Bevers, P., Hartley, M., and Plotkin, A. (2003). The psychosocial impact of closed-circuit television low vision aids. *Visual Impairment Research*, 5, 179–190.

Szlyk, J. P., Seiple, W., Laderman, D. J., Kelsch, R., Stelmack, J., and McMahon, T. (2000). Measuring the effectiveness of bioptic telescopes for persons with central vision loss. *Journal of Rehabilitation Research and Development*, 37, 101–108.

US Department of Education. (2004). *Individuals with Disabilities Education Improvement Act of 2004*. Retrieved March 2, 2009, from http://idea.ed.gov/download/statute.html.

Veraart, C., Duret, F., Brelen, M., Oozeer, M., and Delbeke, J. (2004). Vision rehabilitation in the case of blindness. *Expert Review of Medical Devices*, 1, 139–153.

Verbrugge, L., Rennert, C., and Madans, J. (1997). The greater efficacy of personal and equipment assistance in reducing disability. *American Journal of Public Health*, 87, 384–392.

Virgili, G., and Acosta, R. (2006). Reading aids for adults with low vision. *Cochrane Database of Systematic Reviews*, Issue 4, Art. No. CD003303.

Virgili, G., and Rubin, G. (2006). Orientation and mobility training for adults with low vision. *Cochrane Database of Systematic Reviews*, Issue 3, Art. No. CD003925.

Wilson, D. J. (2005). Braces, wheelchairs, and iron lungs: The paralyzed body and the machinery of rehabilitation in the polio epidemics. *Journal of Medical Humanities*, 26(2–3), 173–190.

World Health Organization. (2001). *International Classification of Functioning, Disability and Health: ICF*. Geneva: WHO.

(2003). *International Statistical Classification of Diseases and Related Health Problems: 10th revision*. Current version. Version for 2003. Chapter VII. H54. Blindness and low vision. Retrieved March 2, 2009, from www.who.int/mediacentre/factsheets/fs282/en.

Wressle, E., and Samuelsson, K. (2004). User satisfaction with mobility assistive devices. *Scandinavian Journal of Occupational Therapy*, 11, 143–150.

9 Mothers of invention? The myth-breaking history and planetary promise of women's key roles in subsistence technology

Rae Lesser Blumberg

Technology is usually depicted as masculine: from stone tools presumably made by men for men's pursuits to men building rockets that reach the moon and beyond. This overwhelming identification of men and technology is one of the reasons so few girls dream of or enter engineering, even today. Enduring a dream denied, or being treated like an outsider in one's chosen field, negatively affects one's psychological well-being. In this chapter, I suggest that the conventional wisdom that has helped deter many women from greater engagement with technology is misleading or wrong.

Instead, I argue that for almost all of human history – and among our nearest nonhuman relatives – we have been closer to equal opportunity in technology, indeed, to the "mothers of invention" model, than the stereotyped notion of "technology=male." I further contend that this has contributed to human well-being, both material and psychological. To support my thesis, I will trace female/male contributions to what Gerhard Lenski has argued since 1966 is the most crucial of all forms of technology: subsistence technology, or how we make a living from our planet. This entails examining my version of Lenski's "modes of subsistence." I call them "techno-economic bases" to highlight the linkage between technology and economy in our past, present, and future.

I posit that humans' "historical main line" has consisted of only four major bases and that a fifth may be emerging in today's globalized world: (1) foraging (hunting and gathering); (2) horticultural (hoe or digging stick cultivation); (3) agrarian (plow-based); then – following the rise of capitalism – (4) industrial. The emergent base is global, but still lacks a final name: the "information," "information/biotechnology," or perhaps "knowledge/life" society. I then factor gender into a Big Picture overview of these techno-economic bases. In preview:

Written with the excellent and gratefully acknowledged assistance of Diana Bowen, University of Virginia.

- Foraging, where women apparently were at least as important as men in the techno-economic base of most non-Arctic groups, began millions of years ago when we were hominids and this basically egalitarian, sharing way of life accounts for nearly all of our history as Homo sapiens, too.
- Women are usually credited with developing cultivation, perhaps the most consequential technological innovation of all time. This led to the world's first horticultural societies a mere 10,000–12,000 years ago. Evidence indicates that female status initially was quite high and that early horticultural societies tended to be peaceful and innovative.
- Then women's position (including as technology innovators) dropped wherever patriarchal, hierarchical, warring societies rose and prevailed. These included many later horticultural societies, most agrarian societies (which first emerged 5,000–6,000 years ago in the Middle East, and where, on average, women's position hit its nadir), and industrial societies over the last two centuries (but where, especially since *c.* 1970, women have been regaining ground).
- Overall, it appears that until *c.* 6,000–7,000 years ago, women in most societies played an essentially equal part in techno-economic pursuits, including control of resources. Levels of gender equality were also generally high.
- Even during these last 6,000–7,000 years, women's eclipse in techno-economic activities and innovation has been far from total or uniform, despite the stereotypes. In at least some groups at each techno-economic level, a less patriarchal model prevailed.
- Now, an information/biotech/globalized new techno-economic base seems to be rising, along with women's importance – and level of equality. In general, the nations where these techno-economic *and* gender transformations have advanced the most have fared the best in enhanced wealth and welfare.

The remainder of the chapter is organized as follows: First, I present a brief look at links between females and subsistence technology among chimps and bonobos, with whom we shared a common ancestor until *c.* 5.5–7 million years ago (De Waal, 2005, uses the 5.5 figure). Second, I take a gendered look at foraging, horticultural, and agrarian techno-economic bases. Third, I summarize my general theory of gender stratification and show how it helps illuminate gender, economy, and technology in any time period or group. Fourth, I give an overview of relevant trends in the US and other industrial societies. The final section explores what is happening now as a new base appears to be coalescing – and creating new patterns of increasing gender equality at its cutting edge – amidst the rising uncertainty and turbulence of an uneven transition and accelerating technological change.

Female chimps and bonobos and subsistence technology

We tend to think of chimpanzees as our closest nonhuman kin. They certainly are both more numerous and studied than our equally close relative, the bonobo. There are *c.* 200,000 chimps in the West African wild, versus only *c.* 20,000 bonobos (mostly in war-torn Eastern Congo) – and *c.* 6.7 billion of us. Here is a brief summary (from De Waal, 2005, unless otherwise noted):

- Among ***chimps***, males are more organized than females. They compete for political power and generally have the upper hand – but it is females who have been observed making and using most of the tools utilized in getting food (i.e., subsistence technology).
  - This includes, for example, poking long sticks into crevices to dig out insects, cracking hard nuts with stones, making sponges by chewing leaves into wads – and teaching these techniques to offspring and others.
  - The newest discovery is somewhat shocking: it was *female* chimps in Senegal who proved to be the main makers and users of **weapons** in **hunting** – long thought to be a disproportionately male activity among chimps and humans alike. In the first evidence of systematic nonhuman use of weapons, it was females and some immature males who gnawed the ends of sticks to create rudimentary spears, which they used in twenty-two observed hunts of bushbabies, a small primate (Henderson, 2007; only one hunt succeeded). Jill Pruetz, the lead researcher of the study, theorizes that female chimps pioneered hunting with weapons because bigger and stronger males – who did not share bushbaby meat – can hunt with their bare hands, but females hunt better if aided by weapons (Henderson, 2007).
- Among ***bonobos***, females are far more organized than males and use their organization to control the food supply; it is females who have the upper hand. Unlike the squabbles and episodes of violence found among chimps, bonobos make love not war: On average, every 1.5 hours, they engage in sexual behavior, including genital–genital rubbing in varied combinations of sex, age, and number. They do this when social friction emerges – or just for fun. Research, however, seems lacking on who does more tool-making among creatures that resemble us even more than chimps (e.g., their walk is more upright,

they have a more human-shaped body, they French kiss and often have face-to-face sex).

Gender and technology before the rise of capitalism and industry

As this section will show, women have been major contributors to subsistence technology during virtually all of our history on earth. But we also shall see that this was *not* the case in any of the societies that became industrialized by the early twentieth century. All the European countries in this category, plus Japan and the US, came from the types of agrarian societies where patriarchy reigned and women made few visible contributions to technology.

Foraging: our longest, most egalitarian techno-economic base

During almost all of our history as hominids and then Homo sapiens we lived by **foraging**, more popularly known as hunting and gathering, even though gathering seems to have accounted for 60–80 percent of the calories except at Arctic latitudes (Fry, 2006; Lee and DeVore, 1968). Gathering is mostly female; hunting is largely male. Therefore, women were, at minimum, equal providers at milder latitudes. Our own type of human, Homo sapiens, seems to have arisen in Africa just under 200,000 years ago. The earliest evidence of any cultivation was *c.* 18,000–20,000 years ago (also in Africa), but it was not until 10,000–12,000 years ago that societies living mainly from cultivation (mostly by women) arose in the "Fertile Crescent" of Mesopotamia, parts of Southeast Asia, and elsewhere. So women's importance in subsistence is undeniable. What about women's role in subsistence *technology*?

Let us start with what Sally Slocum (1975) considers two crucial technological innovations almost certainly made by women that contributed greatly to the survival of our kind during the long era of foraging: the **baby-carrying sling** and the **gathering basket**. Unlike those stone tools on which the "man the technologist" stereotype is based, they were made from perishable materials that have left no archaeological remains (LaFee, 2007). Both, however, led to enhanced women's productivity and more reliable subsistence for the group, thereby contributing to our psychological as well as physical well-being as a species.

Slocum suggests the baby sling probably came first, given that we had no fur to which babies could cling. However, fiber and string led to more than containers for gathering: "From these . . . come snares and fish lines, tethers and leashes, carrying nets, handles and packages, not to mention a way of binding objects together to form more complex tools"

(LaFee, 2007: E8). And women's containers surely extended to those for **water** (Engelman, 2008).[1]

Kathleen Gough (1971) argues that humans' patterns of foraging included periodically going out in search of subsistence and then bringing it back to the campsite, an almost always temporary but crucial "GHQ" (general headquarters). Carcasses of larger animals can be slung over shoulders or a pole, but how do people bring back handfuls of seeds, small insects, or animals, berries, roots, and the like? Basketry is the answer and it is an overwhelmingly female pursuit in human groups (Murdock, 1967; Murdock and Provost, 1973). In short, the development of carrying baskets solved the problem of getting the more reliable and more numerous gathering calories back to that home base (yes; gathering may be less glamorous and its calories less dense or craved than animal protein, but it is more reliable). Richard B. Lee, noted ethnographer of the !Kung, asserts (1984: 40): "The universality of the carrying device and its functional importance among all recent hunter-gatherers has implications for the evolution of human subsistence during the Pleistocene, because a device for carrying vegetable foods would seem to be a prerequisite for human economic and social life (see Lee, 1979, pp. 489–494)."

Regardless of which came first, both baby slings and carrying containers helped humans prevail. Most foraging humans had to be nomadic – at least seasonally – in order to survive. On some occasions, a woman might have left her still-nursing child back at the base camp. (It should not have been hard to find a lactating woman to babysit once the child was beyond infancy and nursing less frequently. After all, studies of contemporary foragers [e.g., the !Kung] find that children typically were nursed to about age four and adults only left camp perhaps two to three days a week to hunt or gather [see below].) Still, there were other times when a woman had to take her child. Without a baby carrier, how could she accommodate that child and still gather effectively? How could she carry both her small child and her fair load of possessions (including technology) to the next campsite without carrying devices?

Moreover, the ingenious baby-sling solution had further consequences those original mother-innovators could not have imagined. For example, it permitted the baby easy access to the breast, and because on-demand nursing increases the ovulation-suppressing effect of breastfeeding

[1] In 1542, Juan Cabrillo found the Chumash, acorn gatherers of coastal S. California, living in permanent settlements of up to 800, making and trading superb baskets and watertight bottles. Most baskets were associated with traditionally female activities (LaFee, 2007). And most authorities also see women as crafting the many **cooking** techniques that multiplied edible calories and changed our history.

(e.g., Frisch, 1978), especially if there is no baby food for, say, nine months, this contributed to women having their babies farther apart and a lower total fertility rate.[2] Indeed, studies of modern foragers (e.g., Australian Aboriginals and the !Kung of Southwest Africa) show spacing of four-plus years and a total fertility rate hovering around replacement level: a little over two children per mother (Birdsell, 1968, makes analogous estimates for the Pleistocene). In turn, this fertility pattern facilitated women's productive role. This was especially important among "pure," but non-Arctic, foraging groups, who do no herding, cultivation, etc. (Kelly, 1995). Wide spacing also reduced child mortality: If a baby arrived before its older sibling was weaned, both were at greater risk. Furthermore, longer intervals between (fewer) births also could have positive consequences for women's autonomy and prospects for equality, not just productivity.

The combination of the baby sling and carrying devices made women formidable creators and wielders of a survival-enhancing subsistence technology. Moreover, although almost surely developed by women (and in many places and times by our early ancestors), both these technological innovations could be used by men. Men can carry gathering baskets, water bottles, or even babies in slings – and do, especially in more gender-egalitarian groups.[3]

[2] I posit two types of factors affecting forager women's wide spacing/low fertility (Blumberg, 1984): (1) *Passive* factors including (a) nursing practices such as on-demand nursing, as well as weaning at about age four, and (b) low body fat ratios (Kolata, 1974) stemming from their nutritious but low-fat diet and their high levels of physical mobility. (2) *Active* factors including (a) infanticide, (b) abortion, and (c) use of plants with contraceptive properties – which seem to have been used for this purpose far back in our history (e.g., Engelman, 2008).

[3] For example, in the course of my work and research in forty-plus countries, I have personal experience with five groups that are gender egalitarian at the local level (provided you do not consider local manifestations of the government and world religion, both of which are more male dominated than local village life). These include the indigenous people of the Central Andes of Ecuador; the people of Northeast Thailand, and their Lao co-ethnics across the Mekong River in the lowlands of Laos; the Bijagos, who live in a string of islands off the coast of Guinea-Bissau in West Africa; and the Mosuo (they call themselves Nazeh) of China's Yunnan Province in the country's southwest. Here is a vignette about a young father and his baby-carrying sling in the wild and wooly frontier town of Oudomxai, Laos, near the border with Yunnan Province, China:

> A young Lao Lum [lowland Lao] man was walking down a side street toward the main road to China. He was carrying a baby, probably close to a year old, in a narrow cloth hip sling suspended from his shoulder. As he approached the corner, a teen-age girl on a motorcycle zoomed into the side street from the road. She was dressed in school uniform: white blouse and a navy blue *sin*, the long, dark traditional embroidered skirt . . . His hand reached down as she whizzed by, and he protectively and tenderly patted the baby's head. Then he began to cross the main road. (Blumberg, 2002)

Any discussion of gender and technology among foragers also must emphasize how both men and women were able to make their own basic tools, which they then subsequently *controlled*. A woman could produce her digging stick (which also could serve as a basic weapon for defense or offense, although war was very rare among "simple" nomadic foragers [Fry, 2006]). A man could fabricate a bow and arrow, and/or spear/spear-thrower (even if some were better or more specialized at it). In other words, each had equal control over the *means* of production. Thanks to the apparently universal custom of **sharing** among foragers (Blumberg, 2005; Lee and DeVore, 1968), both had about equal control over the *fruits* of production. (If anything, rules about sharing scarcer, less reliable animal kills were stricter than rules about sharing the results of routine gathering.) In terms of my gender stratification theory (see below), this equal gender division of the most important techno-economic resources is a primary reason for the high levels of gender egalitarianism among most pure, non-Arctic nomadic foragers.[4]

Women, men, and technology in horticultural societies

The bottom line is that almost all authorities credit women with the slow development of cultivation. That may be the greatest "mothers of invention" story of all time.

First, however, let us clear up a common error: Early cultivation was ***horticulture***, *not agriculture*. Horticulture involves using a hoe or digging stick to cultivate a small garden plot, typically for only one to several seasons, whereas agriculture involves permanent plow cultivation on large fields. Lenski notes that horticulture comes from the Latin word for "garden," whereas agriculture comes from the Latin word for "field."

Why do women get the credit? The short explanation is that men, as the predominant hunters, can be considered the zoology specialists, whereas women, as the main gatherers, had superior botanical knowledge. It is believed that all groups of forager Homo sapiens had figured out that seeds sprout into plants. Their lives depended on a generally nomadic cycle of "seasonality and scheduling" (see Blumberg, 1978). This meant

[4] Australian Aboriginals are an exception. As explained below, they were forced to remain foragers out of necessity, for some 40,000–60,000 years, in fact. But their typically male-tilted kin/property system is as elaborate as in many horticultural societies (family and kin complexity reach their zenith in prestate advanced horticultural groups [Blumberg, 1978]). Middle-aged men could marry multiple wives, whose gathering provided resources for these men to buttress their position. Overall, these groups were less gender egalitarian than most other non-Arctic "pure" foragers. (Arctic groups, with male-dominated production, also tend to be less gender equal.)

being at the right place at the right time to obtain the food resource in question – especially if there was nonhuman competition. This also gave women more reasons for close observation of growing cycles.

But input–output analyses (Lee, 1968, 1969) of the !Kung Bushmen of the harsh Kalahari desert find foragers to be, perhaps, the world's most leisured people (Sahlins, 1968): The average workweek was twenty hours or less, spread over 1.9–3.2 days a week (Lee, 1984: 51). And only about 65 percent of the population – adults *c.* 15–60 – were involved in subsistence pursuits. For untold millennia, it would appear, there was little need for most people to work harder to produce a basic food supply. (Foragers apparently produced surplus mainly for festivals and, in harsh climates, winter. Otherwise, it was too hard to take along surplus when they moved camp so they had little incentive to accumulate it.) It would not have taken long for them to figure out that cultivation entailed more work than foraging.

So, why did they ever leave paradise – the peaceful, egalitarian, leisured bands depicted in most ethnographies of pure foragers (e.g., Marshall, 1965; Turnbull, 1961, 1981)? The gradual rise of **population pressure** is the most agreed-upon reason. All the cultigens (edible plants cultivated for that purpose) that were massaged by our ancestors into today's staple foods – wheat, rice, corn, cassava, quinoa, yams, potatoes, etc. – share the characteristic of being genetically plastic (i.e., mutating easily). Once women felt the need to increase the food supply to feed growing populations, they presumably would have begun selecting for and planting the most food-rich and desirable plants, capitalizing on the most promising mutations.

According to Diamond (1997), all the world's large land masses had at least one such suitable cultigen except for Australia, which had abysmal "geographic luck." (As a result, foraging remained the sole techno-economic base there until the British sailed into Botany Bay in 1788.) Everywhere else, in a remarkably short time, early experiments in cultivation – presumably done by the women, who dominated gathering – gradually led to an increasing reliance on horticulture, which permitted more settling down. As a result, by *c.* 10,000–12,000 years ago, we can talk about horticultural *societies* and the beginning of the "horticultural era" (Nolan and Lenski, 2009).

Today, most horticultural societies are "female farming" systems (e.g., Baumann, 1928; Boserup, 1970). Women not only are credited with developing what is arguably the world's most revolutionary technological innovation, they now are primary (or, at least, equal) farmers in four-fifths or more of world horticultural societies (Bryson, 1981). These days, however, most groups that continue to practice this cultivation method do so because they live in areas with poor, acidic tropical soils too thin for the plow – the situation in much of sub-Saharan Africa, where women still raise up to 80 percent of locally grown food crops (Saito and Weidemann, 1990).

Technologically, horticultural cultivation can be done with the same digging stick used for foraging, but attaching a blade to the stick – that is, turning it into a hoe – raises its efficiency. Before metallurgy, a mostly male pursuit, women easily could have made the wooden blades and bindings for hoes. Most processing equipment would have been similar to items already in use for the products of gathering. But horticulture soon would have produced so much higher yields that needed processing and storage that women's work would have gone up (Engelman, 2008: 88–89).

Men's work also would have been affected. In forested areas, the main horticultural technique is slash and burn, or shifting, cultivation. At the end of the dry season, the bush (but usually not the top canopy trees) is cut – usually by males – and allowed to dry. Then they do a carefully controlled burn. This releases all the nutrients in the plants into the ashes. As a result, in the first season, whatever is planted tends to produce high yields with very few weeds. There is a catch: If a plot is used too long, weeds multiply, yields drop, and cultivators' labors intensify. For most horticulturalists, it is time to move on. Men will gain fresh hunting grounds, even if they have to do most of the clearing work for the new plots. Women will have far less weeding.

Most important, however, is that with cultivation people became semisedentary **villagers**. They moved the village only after a number of years, when suitable nearby plots had been used and "commuting" to increasingly remote gardens or good hunting areas loomed as an issue. The new settlement pattern permitted higher fertility as women moved about less and became a little plumper, thereby facilitating pregnancy. More settled people would see less reason to kill a baby born "too soon" after the last birth (say, three years). The result was the world's first "population explosion," which, in turn, precluded a return to foraging without a major die-off. The more settled village life also permitted utilization of two technologies that already had been invented, but were little used because they involved items too bulky to carry around on the at least partially nomadic "seasonality and scheduling" rounds of foragers. These were: (1) **pottery**, and (2) **weaving**.

Both contributed greatly to our physical and psychological well-being. Pottery expanded the range of food preparation, presentation, and preservation: It permitted many more dietary and storage options than the throw-away "plates" (of leaves, etc.) used by many foraging groups and their basketry or animal intestine or hide containers for keeping food and water. Weaving brought the greater comfort, weather appropriateness, and health benefits of better-made and better-adapted garments. Again, the ethnographic data bases, including Murdock's *Ethnographic Atlas* (1967; Murdock and Provost, 1973), show these to be largely *women*'s activities.

In sum, women proved both the "mothers of cultivation" and main farmers in most such groups, as well as the key creators/users of the two major new technologies linked to horticultural village life. Then and now, horticulture is considered the most feminized techno-economic base.

However, as discussed in the gender stratification section below, mere work in key subsistence techno-economic activities is *not* enough to ensure high levels of equality for women. While some work in key productive activities seems to be a necessary prerequisite for greater gender equality, it clearly is insufficient. Only where work leads to control of technology and/or economic resources (e.g., income) does it translate into economic power, which I term *the* major "poker chip of power" for women (Blumberg, 1984).

The extent to which women have economic power and/or control technology varies widely in horticultural societies, depending on the **kinship/property** system. Also, where **war often comes close to home**, men are provided with an additional source of resources and technology while women's ability to make a livelihood and overall position are constrained.[5]

Matri-oriented societies The Iroquois of colonial North America constitute one of the best known examples of a horticultural group where women controlled both the economy and horticultural technology. Their more patriarchal American colonist, British, and French neighbors even considered them a "matriarchy." That is an overstatement. On the one hand, the kinship/property system *was* female dominated: Descent was matrilineal and, when the young couple married, they went to live in a longhouse with the bride's female kin. Moreover, Judith Brown (1975) describes how the women exercised almost total techno-economic control. They not only dominated cultivation, but also controlled land rights and owned the stored crops. On the other hand, men predominated in the military and political spheres. Nonetheless, women had some say in these areas, too. If the women (especially the

[5] Collins long has stressed this factor (e.g., 1971, 1975; Collins *et al.*, 1993; Coltrane and Collins, 2001).

Council of Matrons, composed of the senior women who headed each longhouse) opposed a particular warfare expedition, they could refuse to give men the trail food they would need, thus exercising military veto power. They also had influence over the political sphere: Women from the Council of Matrons would nominate a man to be chief and recall him if they were unsatisfied with his performance, nominating another man in his stead. All in all, however, the female-controlled economic sphere and the male-dominated military and political spheres probably balanced out to a more or less "50–50" partnership arrangement.[6]

Patri-oriented societies Most surviving horticultural groups today are in sub-Saharan Africa and fully 75 percent are patrilineal/patrilocal (Eloundou-Enyegue and Calves, 2006). Women in these groups may be the principal farmers, but they lack de facto land rights. Laws giving women rights to inherit land that have recently been passed by various African countries are still rarely enforced in rural areas (Blumberg, 2004b). So, do women have ways of transforming their labor in cultivation into control of income? This varies by region. In West Africa, vibrant local markets dominated by women sellers have a long history. For example, they apparently were flourishing in Nigeria *c.* six hundred years ago (Sudarkasa, 1973). In East Africa, however, such markets have been far less developed and women were less likely to be traders, so they were less able to translate their farm labor into economic power. As a result, they have had much less autonomy and remain less able to decide on – and purchase – the technology they use in cultivation. However, all over sub-Saharan Africa – West, East, or South – female farming is low-resource farming, so the technology level remains low (Blumberg, 1992).

[6] Blumberg (1978) describes "four paths through horticulture." One path remained quite egalitarian. The other three involved complex unilineal kin systems and the pursuit of surplus to varying degrees. Historically, most matrilineal horticultural societies remained internally peaceful, productive, and monogamous. But ultimately most were eclipsed or conquered by patrilineal societies. These patri-oriented groups' male leaders could concentrate more resources at the top because they tended to have multiple farmer-wives who produced surplus that could fund their political gambits and the occasional external or internal war. But the patri groups that became the "historical main line" avoided the dead end of another patri path: "warrior complex" societies so trapped in feuding/revenge warfare (often over women, who were made scarcer by female infanticide) they rarely developed the state. (To add a personal note, I have done fieldwork in all three unilineal types. As a Peace Corps Volunteer, I led an expedition to the headwaters of Venezuela's Orinoco River to study the matri-oriented Maquiritare and the notorious "warrior complex" Yanomamo [Blumberg, 1978]. Since then, I have done development work/research with various groups [overwhelmingly patri-oriented] in ten African countries and a few Amazonian tribes [two of them matri-oriented] in Ecuador and Peru.)

There is one dramatic difference in African horticultural societies with male- versus female-oriented kinship/property systems. In the former, what Murdock (1967) terms "general polygyny" (where at least 20 percent of unions involve more than one wife) remains common (United Nations, 1987). Women's productive role and expertise in horticultural techniques and technology are the main reasons that, in groups where the bride goes to live in the husband's village, most require that "bride price" be paid by the groom and his family to the bride's family and the bride. This reflects the woman's value as a creator of both subsistence and surplus (Clignet's [1970] classic study of polygyny is titled, *Many Wives, Many Powers*).

In my own fieldwork, I have found one scenario where women who are key farmers in a patrilineal/patrilocal group have considerable control over subsistence technology and a strongly innovative role in improving that technology: *The prerequisite is that women have a stake in the outcome.* This means that they have economic autonomy, backed, too, by traditions of female *organization* that guarantee that a woman who increases her income through technical innovation will not see it confiscated by her husband. This is just what I found among the Igbo of Southeastern Nigeria, who fulfill all these criteria. The Igbo are a "female farming" group that still practices polygyny and has bride price, despite being overwhelmingly Christian.

Igbo women have been able to push horticultural cultivation techniques and technology to unusually high levels. I worked with agricultural economists who studied the "compound gardens" maintained by these women behind their homes. There, they grew up to seventy to seventy-five cultigens in a marvelously sophisticated quadruple canopy system that had as its objective, "something to eat, something to sell, every day." They also grew medicinal and ornamental plants. They used taller species to shade lower-growing, more delicate ones; planted nitrogen-fixing species to fertilize others; used some plants as natural insecticides to protect their neighbors; fertilized with household organic refuse; and irrigated where necessary with used household "gray water." They actually achieved higher yields with their plantains than agronomists at the University of Nigeria-Nsukka (Blumberg, 1986). And they achieved permanent cultivation in those gardens. Also, these women had true technological expertise with respect to chemicals, fertilizers, and what worked best with each major crop (one junior wife I interviewed, who described herself as the principal farmer of the household, astounded the agriculture grad student who was interpreting for me with the extent and specificity of her knowledge).

In sum, in a case where women retained *incentives* (these Igbo women, buttressed by their in-marrying wives' organizations, controlled the proceeds from their compound garden sales), they raised horticultural techniques and technology to significantly better levels. They even bested trained male professionals in the case of plantains. What happens to innovation when producers cannot count on incentives? To answer that question, we have to turn to the mostly oppressive, patriarchal, and warlike world of agrarian societies, where even male peasants are more likely to be economically powerless serfs than freeholders.

Bad times in the average agrarian group and why the average is misleading

Akin to Dickens's *A Tale of Two Cities*, in the "average agrarian society" it was the best of times (for the agrarian ruling class) and it was the worst of times (for the common peasants and, especially, for most women).[7] But averages can bury important information.

In Blumberg (2004a), I posit three different types of agrarian societies, which differ in the gender division of labor in subsistence technology, as well as in the gender division of *control* of economic (and technological) resources (below I link this to my gender stratification theory).

First, instead of Lenski's technological criterion (the presence of iron) to separate agrarian societies into "simple" and "advanced," I used a different technological criterion: rain-fed versus irrigated cultivation ("dry" versus "wet" – with all the major "wet" agrarian societies growing irrigated rice as their principal crop). This clearly distinguishes two very distinct gender divisions of labor: (1) Rain-fed agriculture is the world's most *masculinized* techno-economic base (Boserup, 1970, calls these "male farming systems"). (2) In contrast, irrigated cultivation of rice is so labor-intensive that *everyone* works: men, women, boys, and girls, and available major domestic animals such as oxen, water buffalo, donkeys, and horses.

Second, I then divided the irrigated societies on the basis of their kinship/property system. The great majority was *patriarchal*, but a

[7] Agrarian societies are based on plow cultivation and some background is useful here. Someone invented the plow somewhere in the Middle East 5,000–6,000 years ago. Technically, it digs deep into the soil, bringing up nutrients and breaking up weeds (Nolan and Lenski, 2006). The Eurasian landmass had superlative "geographic luck," for example all the most valuable domesticatable animals (Diamond, 1997). So, farming systems were mixed and animals provided free fertilizer. All this permitted permanent cultivation and settlement and higher levels of surplus/stratification/hierarchy/warfare (Lenski, 1966). As the plow diffused along Eurasia's long East–West axis, subjugation entered as well.

minority – all in Southeast Asia – had gender-neutral to female-favoring systems. This produced a clear-cut differentiation of the gender division of *resources*, with women faring far worse in the patri-oriented type.

Patriarchal agrarian societies probably accounted for all but a small percentage of the population of all agrarian societies prior to the industrial era. On average, in both dry and wet types, the position of women – and peasants, who made up *c.* 80–90 percent of the population – was generally unenviable. (In such groups, the rate of innovation in subsistence technology *slows* because those who have the technical knowledge have no incentives – or access to the needed resources – to improve productivity. Lenski coined the word "technostasis" to describe this.)

What about the "nonpatri-oriented minority"? The current population of the Southeast Asian countries where this system is prevalent is roughly half a billion, about the same as the European Union. Even so, they still make up well under 20 percent of the population of contemporary "industrializing agrarian" societies (Nolan and Lenski, 2006). In fact, just two of the patri-oriented industrializing agrarian nations, India and China, account for close to 2.5 billion people, nearly 40 percent of the total global population.

The net result is that any average for agrarian societies – traditional or industrializing – is going to totally obscure the far more egalitarian patterns of technology and economy as well as gender stratification found within most Southeast Asian societies, especially in wet rice villages. Therefore, it is worth considering these peoples separately here.

Southeast Asian bilateral/matri-oriented irrigated rice societies

- They typically have less socioeconomic stratification at the village level than the two patriarchal types: Geertz (1963) describes the villages of mainly Muslim Java, which contains over two-thirds of Indonesia's current population of *c.* 232 million, as divided into the "just enoughs" and the "not quite enoughs." Villages in Buddhist Northeast Thailand, which contains about half of Thailand's current population of *c.* 66 million, are often similar (Blumberg, 2001a). Tellingly, in both places, women rarely are economically dependent on men.
- Subsistence technologies tend toward small-scale irrigation projects that do not rely on distant governments to build and (fail to) maintain. Rather, the more equal water allocation spurs men and women alike out of bed to repair a break in the system, even in pouring rain (Blumberg, 1981).
- And they are likely to have small-scale and ingenious technology for other aspects of water management – including water storage in the dry

season. In Northeast Thailand and lowland Laos, I saw how these not only facilitated dry-season crop production, but also made women's daily burdens easier because water was available for both household and productive use.

- Furthermore, women tend to be entrepreneurs who have developed technologies to use water for profit-making ventures (e.g., making batiks). This is part of the strong female economic traditions found in these societies, even the Muslim ones.[8] Many of these women not only trade in the market, but also create small-scale, technology-based microenterprises.

The Southeast Asian pattern has predominated in, for example, Thailand, Burma, Laos, Cambodia, Indonesia, and the Philippines. It is generally less hierarchical, centralized, and male-dominated than the two more patriarchal types of agrarian societies.

Patri-oriented agrarian societies

- Concerning the gender division of labor, there is little difference between the Southeast Asian irrigated rice societies and their more patriarchal counterparts in, for example, Taiwan, and the southern parts of China, Japan, India, and Korea. But, as noted, there is a big difference between the irrigated and rain-fed types of agrarian societies, with women far more involved in production in most wet than dry agrarian groups. In both wet and dry *patri*-oriented types, however, women usually have a low level of *control* of economic resources and technological tools; they are essentially serfs on land controlled by their husband and/or his male kin.
- Even here, however, the level of gender stratification is not uniform. Nor is women's relative degree of involvement in and control of subsistence technology. For example, during the Viking era, Scandinavian men often left for long periods, even years. Women then played key roles in the rain-fed farming system and in local business, acquiring and inheriting property. And the overall level of patriarchy was markedly lower than in, say, the rain-fed agrarian empires of the Middle East (e.g., of Persia and the Ottoman Turks). More commonly, however, frequent warfare close to home further depressed the poor overall position of women in most patriarchal agrarian societies.

[8] Gonsoulin (2005) researched matrilineal, quite gender-egalitarian Muslim peoples (e.g., the Minangkebau) in Southeast Asia. There, Islam was brought by traders, not the sword, and did not include *sharia*, just the Koran. So *adat*, customary law, survived, and women retained equal inheritance, own-account trade in the market, etc.

Not surprisingly, it is in rain-fed agrarian societies where there is the greatest prevalence of the custom of dowry, the opposite of bride price. Dowry systems typically are found where women do *not* play an important role in production. In such groups, the bride's family must pay the groom and his family to take her. Dowry has become a big problem in India, especially Northern India. There, dowry demands are reaching new heights not among rain-fed agrarian farmers, but in urban middle classes eager for consumer durables as the economy expands. Excessive dowry demands, and the availability of cheap sonograms (that permit ascertaining the sex of the fetus) and abortions, are further depressing India's traditionally skewed sex ratio (Kishor, 1993). In fact, Harriss-White (1999) argues, frequent female feticide and elimination of unwanted daughters via selective childrearing strategies amount to "gender cleansing." In the countryside, however, dowry continues to reflect women's low involvement in the largely male farming system. Concomitantly, women's role in cutting-edge agricultural technology in such systems (e.g., the Green Revolution that has so enriched Punjab and other Northern Indian states) has been negligible and they often still are viewed as economic parasites and technologically uninformed.

In conclusion, in our survey of history we have seen that only in the two patriarchal types of agrarian societies has technology been a male preserve, but *all* of today's advanced industrial societies came from these two types.[9] Little wonder that the notion of "mothers of invention" seems so strange to us.

A few relevant hypotheses from my gender stratification theory

As background, among humans, the range of gender stratification is from extreme patriarchy up to "50–50" partnership. There are many societies where male hegemony is very strong and the position of women very subordinated, but there is no evidence of any human society where women are hegemonic and subjugate men.

My general theory (e.g., Blumberg, 1984, 1991, 2004a) posits many variables that affect the level of gender equality/inequality in human groups, but the most important one is hypothesized to be relative **economic power** – relative control of economic resources, especially surplus, by males versus females in "nested" levels ranging from the micro (the couple) to the macro (the state).

[9] The story of how we made the revolutionary transition from relatively stagnant agrarian societies to dynamic industrial societies involves the rise of capitalism, the "once in a planet's history" discovery of the New World, with its gold and silver, and many other factors. This topic is beyond the scope of the chapter (but see Blumberg, 1978).

Interestingly, of all the major sources of power I treat in my theory (economic, political, force, ideology, and information), only **economic** power goes above the "50–50 line" for women. There are societies where men control almost the whole economy but, as among the Iroquois, in other groups it is women who control the economy "lock, stock, and barrel." Empirically, in fact, economic power can range from near-zero to near-total for both genders.

(My theory even is germane in helping to explain why bonobos are the only apes among whom females have the upper hand. The bonobo habitat is lush and food is easily obtained. Females migrate into the troop at adolescence, as among chimps and gorillas, too. Only among bonobos, however, are they *organized* and, more important, only among bonobos do females *control the food supply* – that is, have the economic power.)

Concerning the other main types of power, **ideologies** range from male supremacist to egalitarian, but no known group espouses female supremacy. Nor do we know of societies where women have been above the "50–50 line" for *either* overall **political** or **information** power.[10] And with respect to the power of **force**, women are far more likely to be its victims than its wielders (here, men's one-third to one-half more upper body strength and long-standing control of heavy weapons clearly are factors).

In a nutshell, I propose that women's economic power is the most important and achievable route to greater gender equality. How, then, do women gain economic power? A necessary but insufficient precondition, I hypothesize, is their involvement in key techno-economic production activities. "Mere work" in production, however, does not lead automatically to economic power (consider all the work done by the world's slaves, workers, and peasants). Rather, for women, I suggest, it is more like paying the toll to get onto the road leading to greater gender equality. Economic power is what drives them farthest and fastest down that road. Therefore, women's productive labor must be translated into economic power to reap the benefits of greater equality. How? I posit three sets of variables that affect the extent to which women's work in production can be transformed into economic power. The first is the "strategic indispensability" of a labor force.[11] My theory proposes six main strategic indispensability factors, including relative control of technology/technical expertise (another one is the extent to which a group is organized, as noted above concerning Igbo women and female bonobos).

[10] But women gained over half the seats in Rwanda's lower house in mid-September 2008, a world first, even though the state remains largely male-led.

[11] The two other sets of factors are the kinship/property system, and the level of inequality in a society's overarching socioeconomic stratification system (Blumberg, 1984).

Women market traders in Ghana provide an example of strategic control of technical expertise. Except for the top levels of international trade, the marketing/distribution system is mostly in female hands. In the 1980s, women controlled the system's techniques, information, and organization so thoroughly that when the Jerry Rawlings administration tried to break the women traders' power by bulldozing the main market in Accra and executing some top women entrepreneurs for "profiteering," the effort failed. Those government men just did not have the technical knowledge to run the system.

To reiterate, if neither "mere work" nor "mere technical/technological expertise" can be turned into control over resources, they do not lead to more economic power or its key outcome: greater equality. Conversely, I posit that those with greater economic power – male or female – tend to have greater control of their work and technology, and more incentive to innovate in the areas most important for their livelihoods.

How do I measure greater equality? My main **dependent variables** involve control of one's destiny, through "life options" that occur in all human societies, for example marriage, divorce, sex, fertility, freedom of movement, and household power. For instance, concerning marriage, *whether*, *when*, and *with whom* are prime considerations. With greater economic power, women can increase their say in this major determinant of their life course, happiness, and psychological well-being. I explicitly propose that work that does *not* lead to control of economic resources/surplus does *not* lead to enhanced life options, whereas greater economic power *does* result in more say about one's person and destiny.

Three other aspects of my theory also bear mention here. These, too, are outcomes of greater female control of economic resources. First, with greater economic power, one quickly gains greater self-confidence, the foundation of psychological well-being. Soon, this usually leads to more household power, with more "voice and vote" in family decisions. These include more say about fertility, domestic well-being (e.g., when is the baby sick enough to be taken to the clinic and does this differ for boy and girl babies?), and economic issues (acquiring, allocating, and/or disposing of assets). For women, greater say in fertility is especially important, given its disproportionate impact on female life chances. With greater economic power, a woman has more leverage to achieve fertility patterns that correspond to *her* preferences and utilities (e.g., concerning age of first birth, age of last birth, spacing interval between births, total number of children, sex ratio, and the means used to achieve these outcomes).

Second, I posit that women and men with provider responsibilities typically spend income differently: *Women tend to spend income they control more single-mindedly – and even-handedly – on the well-being of their sons and daughters*, and to hold back less for themselves. So a woman who controls income not only has more say in household decisions about her children's human capital, but she can also use her own income to invest directly in her children's welfare: their education, health, and nutrition (Blumberg, 1988, 1989a, 1989b, 1991, 1993, 2009, as well as Blumberg *et al.*, 1992, show mounting evidence for my hypotheses). Those healthier and better-educated girls and boys have better personal futures and also contribute to greater techno-economic development, national income growth, and human welfare in their societies. I call such micro and macro effects the "synergy bonus" of women's economic empowerment (Blumberg, 1989a).

Third, where women have *consolidated* economic power, they tend *not* to be victims of violence (Levinson, 1989, in a ninety-society study, found female economic dependence the main predictor of male violence against women [VAW]; I found similar results in a sixty-one-society study [Blumberg, 1978]). The dark side, however, is the more men feel threatened by women's rising relative economic position, the more likely a short-term spike in VAW (Blumberg, 2004c).

All these points are relevant because with the rise of industrial societies, an initially small but growing proportion of women began to earn and control income. Now, this trend is worldwide and accelerating (the proportion of women earning income is growing faster than among men) as both the cash nexus and what seems to be an emerging "Base V" extend their global reach. At this juncture, let us consider industrial societies.

Industrial societies: toward a historic reversal for women?

By 1750, the Industrial Revolution was in full swing in England, which became the first industrial society by *c.* 1800; starting with Germany and France, other nations soon followed. Lenski (1966) has written of one historic reversal in industrial societies: For the first time since surplus began to be systematically accumulated in horticultural and, especially, the two patriarchal types of agrarian societies, the level of internal economic inequality began to *drop* – there is less inequality within the average industrial society than within the average agrarian one. Moreover, in industrial societies, there are fewer constraints to technological innovations and more incentives to create them, partly because, for more people, there is a clearer connection to economic reward.

Another major trend in industrial societies, from the early twentieth century onwards, has been the incorporation of increasing proportions of women into the labor force. At the micro level, this has led to more equality in the home, as indicated in the theory section above, as well as reduced depression among married women (Coltrane and Collins, 2001). At the macro level, it has also led to greater national income growth (Goldin, 1986). In fact, the *Economist* (2006a: 16) dramatically concludes that the "increase in female employment in the rich world has been the main driving force of growth in the past couple of decades. These women have contributed more to global GDP growth than have either new technology or the new giants, China and India." And, in India, for that matter, the states with the highest proportion of women in the labor force have grown the fastest; additionally, they are the ones that have had the biggest reductions in poverty, according to the World Bank (Faiola, 2008: A12).

We can trace the path of women's employment in the United States. In 1900, only some 20 percent of women in prime labor force years, 18–64, were in the labor force. By 1970, female labor force participation (LFP) had risen to 50 percent (Oppenheimer, 1973), and the overall upward trend has continued since. By the late 1990s, almost 80 percent of married women with children, living with their husbands, were in the workforce – and more egalitarian "sharing" marriages, where both partners help provide and care for the household and children, had become the norm (Coltrane and Collins, 2001). Today, women are approaching 50 percent of the total US labor force. What is arguably more important is that, by the year 1900, women had become sex segregated into the two categories of jobs destined to grow the most ever since: clerical and service occupations. As further discussed in the next section, an increasing proportion of those jobs involve information and life (e.g., health, biology).

Simply put, by the twentieth century, another historic reversal had begun – this one for women. Their growing role in key productive activities – now done for income – started to reduce the inequality they experienced in both the wet and dry patriarchal types of agrarian societies (from which all of today's advanced industrial societies have sprung).

- By the twenty-first century, advanced industrial economies had eliminated most of the nonfarm productive activities performed as *unpaid* family labor, that is, activities that were "mere work," and did not bring economic power to the worker.
- The "second shift" of housework and child care has been more resistant to change, however (Fenstermaker *et al.*, 1991). Even so, since the 1960s, men's contributions to child care tripled, while their housework contributions doubled (Fisher *et al.*, 2006; Sullivan and Coltrane, 2008). This is especially likely where the woman earns a

significant proportion of family income (Blair-Loy, 2003; Blumberg and Coleman, 1989; Blumstein and Schwartz, 1991).

- In the Third World, as the money economy spreads, more people are earning income but, as noted, the proportion of women doing so is rising faster than the proportion of men (Blumberg, 1995). Women who earn and control income are using it to help improve their children's lives and their own relative power and position in the household and beyond. This is so even if women's income comes from the informal sector (as is disproportionately the case in developing countries; in most, the informal sector provides lower earnings, few or no benefits, and low-level technology – but is the fastest-growing sector of the economy).

All these trends – from women's rates of income-earning rising faster than men's, to the increase in dual income couples with more egalitarian and psychologically rewarding relationships, to a now accelerating thaw in the gender division of domestic labor (once thought to be frozen in a very traditional pattern)[12] – are found to differing degrees in every world region.

Helping to disseminate these "patterns of reversal" has been the rapid spread of industrial technology beyond the major centers of northwest Europe, the US, and Japan. Nolan and Lenski (2009) note that, in 1888, only five countries – the US, Britain, Germany, France, and Russia, in rank order – produced 83 percent of world industrial output. As late as 1937, these five produced 78 percent. However, by 2003, these five produced only 41 percent; Japan ranked second and China was the world's third largest economy (2009: 203–204). In addition, women's (paid) labor force participation is relatively high in all the top five. Nevertheless, "culturally male identity" continues to be linked with technical prowess in advanced industrial societies (Wajcman, 1991).

For both genders, however, a shift toward something *post*industrial seems to be simultaneously speeding up and spreading. We appear to be moving (at different speeds in different regions) toward a new globalized world whose lead sector is looking ever more like the fourth change of the techno-economic base in human history. This is potentially earthshaking: Each time the base has changed in the past, it led to transformations in every societal institution from religion, government, law, and the economy to the intimate world of the family – and the relative position of men and women.

[12] For women, for most of the last two hundred years, technological "progress" in housework failed to reduce the time burden involved (Cowan, 1983), or raise feelings of well-being in doing it; here, too, there has been some recent progress.

Base V: will women's improving position help all of humanity?

Women and the new techno-economic base of knowledge and life

To recap, by 1900, women had become sex segregated into the two types of jobs that have grown the most ever since – clerical and service – in our evolving economy. Now, much of our largely service economy is based on information. Take financial services, for example. Trillions are in constant turnover in international currency trading transactions alone. Few of those trades require moving bags of bullion; rather, most involve a few keystrokes on a computer. Moreover, especially at the lower and middle levels, the people working in the resulting finance/financial services system are likely to be women.

Lenski (2005) regards the information economy, from computers to currency transfers, to be only a fourth phase of industrial society. Toffler (1990) and others, however, consider it an entirely new type.[13] Toffler's *Powershift* explicates information as a new, higher type of power. He sees it as more flexible than the lowest grade of power – force. It also is more flexible than economic power. Although money is fungible, printing too much produces runaway inflation, misery, and political instability. However, information is infinite and it is not a zero-sum game. To date, despite efforts by autocratic governments to suppress the Internet and other new forms of communication (e.g., faxes; cell phones with text messaging and photo capacity), most often the technological mice have stayed ahead of the government cats. Information is available – and popularly generated, for example consider Wikipedia – as never before. Moreover, whereas women increased their involvement in the economies and technologies of industrial society only slowly, they are quickly coming to the forefront in two of the main domains of the emerging Base V world.

Not only are most information/knowledge-related jobs still more likely to wear a pink ribbon (female) than a blue ribbon (male) in labor force composition, but with respect to the "techno-economy of life" – life sciences, biotechnology, and health – women's presence is growing by leaps and bounds. According to Burger *et al.* (2007), of all the fields of science, engineering, and technology (SET), it is in the biological sciences that women have made the greatest strides. And it is precisely the biological/life sciences that are ever more important in the emerging,

[13] I clearly lean toward the "new type" view. But even if the blossoming of what I am calling the trees of knowledge (information) and life (biotechnology, biomedical, life/health sciences) proves to be only another phase of industrial society, my basic argument about women's rising role in its globalizing economy and technology is mostly unaffected.

globalized Base V. We have deciphered the information encoded in the human genome and are fast doing the same for species at all levels of evolutionary complexity. Indeed, when the Base V baby is given a definitive name, it may not be just "the information society," but maybe the "information/biotech society," or even the "knowledge/life society," with or without "globalized" as a middle name.

Burger *et al.* (2007: 260) document that, in high school, females take biology courses, as well as advanced placement courses, at a slightly higher rate than males, and women are more than half the undergraduates in the biological sciences. Concerning PhDs, they note:

> Looking only at U.S. citizens and permanent residents, the number of women earning doctorates in the biological sciences (which includes not only botany, immunology, microbiology, and zoology, but also many other fields) increased 631% [from] 1966 to 2004 . . . compared with an increase of only 44% for men . . . women are nearing parity in the biological sciences, receiving 47.7% of the . . . doctorates earned in 2004 (NSF, 2006).

The upside is a flowering of the trees of knowledge and life. In the brave new world of biology, we are poised to turn our exploding knowledge into new global industries and advances in human health, including both physical and psychological well-being. Naturally, there is a downside in all this. Some of those new biotechnologies could run amok, with devastating consequences. For example, Wajcman (2000: 458) argues that now even bodies are made and remade, becoming technological artifacts. Unquestionably, this has risk as well as promise. Alternatively, the forces of reaction and terror could kill the globalized Base V baby while it is still in its cradle.

Barring catastrophe, however, the basic trend lines for both a more gender-equal division of technological expertise and overall gender equality have turned up, although there are significant exceptions. Even now, feminist scientists often feel that SET is still a "boy's club" that marginalizes them in myriad ways – although it is difficult for many women to protest, since they are just gaining a toehold.[14] Still, in an information techno-economic base, education is critical. How have US women fared in SET fields other than biology? Burger *et al.* (2007: 256) detail that:

> chemistry has grown from less than 10% female bachelor's-level graduates in the 1960s to 50% in 2005; physics [noted for being the least hospitable of the sciences to US women] and engineering lag at 25% and 22%, respectively; and

[14] For example, Hubbard (2001) and many other articles in Wyer *et al.* (2001) make a provocative case for women's marginalization, with numerous examples. Tang (2006) traces the invisibility and isolation of most of the small number of women science pioneers who managed to break through.

the percentage of women in computer science has actually declined from 28.6% in 1994 to 27.6% in 2001 (NSF, 2004). However, the lagging fields are still two or three times as popular with women (in total numbers) than they were 30 years ago (NSF, 2000, 2004).

In the US today, the position of women in computer and information technology fields is mixed. On the one hand, teenage girls are now using computers and the Internet at rates similar to their male peers. On the other hand, they are five times less likely to consider pursuing a technology-related career or even taking postsecondary technology classes (Melymuka, 2001). In addition, concerning high school advanced placement exams, Burger *et al.* (2007: 56) state that "the number of females who opt to take the examinations in . . . computer science (1/1000 12th graders) lags behind the number of males (5/1000)."

With respect to undergraduate education, the pattern of women's degrees in computer sciences goes against the gains in other fields: women's share of bachelor's degrees in the computer and information sciences increased steadily through the mid-1980s, when women earned 37 percent of the 38,878 degrees conferred in 1984–1985. Though the *number* of such degrees earned by women continues to rise, the *percentage* was down to 22 percent in 2005. At the master's level, however, the proportion of women receiving degrees increased smoothly, rising to 34 percent by 2000–2001 (NCES, 2002).

With respect to employment, women are about 30 percent of the computer and information sciences workforce (ITAA, 2005; cited in Burger *et al.*, 2007: 265), a higher proportion than in, say, engineering. At the upper levels, women are a third of managers of IT support services or network maintenance teams (ITAA, 2005; cited in Burger *et al.*, 2007: 265). In lower-level jobs, however, women are so overrepresented that these are referred to as "pink-collar ghettos."

To summarize, women are a clear majority in the routine information-handling jobs, but their numbers dwindle in the most technical of the computer/IT fields (see also Cohoon and Aspray, 2006). This is *not* the case, though, in the life/biological sciences, arguably at least as important for the emerging Base V.

What will this mean for women, technology, and well-being? Feminist approaches to technology, per Wajcman (2000: 450), usually stress how it maintains gender hierarchies, not how gender relations may be transformed by new technology. IT may be moving in the conservative direction, reproducing gender hierarchies (though that may be due to a too-narrow definition of what technology-creating jobs are in IT, one that is limited to hardware and software, thereby excluding creative applications). Conversely, the life/biology fields may be more woman-empowering and even transformative, perhaps embodying the positive potential Haraway envisions that science can have on women's well-being (1985, 1991).

Conclusions

As what appears to be the fifth techno-economic base on the historical main line takes shape, we see a number of contradictory trends that complicate any easy conclusions or predictions. On the one hand, the rate of technological change has been accelerating faster and faster, especially computer processing power.[15] On the other hand, the level of uncertainty[16] has been rising rapidly, too (although, fortunately, not as fast).

In today's world, women are 47 percent of members of parliament in Sweden (IPU, 2008) and have just taken over half the seats in the lower house in Rwanda (BBC News, 2008), but Saudi Arabian women cannot drive or vote and, if the Taliban regained power in Afghanistan, they likely would again throw all the girls out of school and all the women out of jobs. There are bitter debates over the costs versus benefits of both globalization and different ways of attacking global warming and/or its consequences. Growing terrorist networks, asymmetric warfare, and epidemics (as well as their media coverage) also increase uncertainty.

At the same time, other, more positive global trends are at work. One important trend concerns the increasing support for women's equality in the world's most global network: the UN system. Very significantly, the third of the eight Millennium Development Goals adopted in 2000 by 189 United Nations member states, to be achieved by 2015, is "promote gender equality and empower women." The United Nations has also adopted gender equality as a *human right*, and it keeps expanding the scope of women's human rights (Blumberg, 1998; Blumberg and Salazar,

[15] Moore's Law, originally stated in terms of the number of transistors that can be inexpensively placed on an integrated circuit, currently predicts, in essence, that the power of computers per unit cost ("bang per buck") doubles every two years (Wikipedia, 2007). It has become a self-fulfilling prophecy that helps drive Base V into the future. The first general-purpose digital computer (ENIAC) occupied 1,500 square feet, weighed 30 tons, and performed 5,000 additions/second; supercomputers now perform over 100 *trillion* additions/second (Nolan and Lenski, 2009: 203).

[16] Uncertainty is usually contrasted with risk. Risk can be calculated if one knows all the values a variable, x, can assume and their respective probabilities of occurrence. Uncertainty is analogous to erasing some of the entries on the risk prediction table: this means not knowing all possible outcomes of a new event or the likelihood of many of the ones one does know about. It is as though someone keeps adding new jokers to the deck: the fall of the Berlin Wall on November 9, 1989, the catastrophe of 9/11 . . .

2006); for example it has banned rape as a war crime. So, there is a new push to reach the "50–50 line" of gender equality in ways only a minority of horticultural and irrigated rice Southeast Asian agrarian groups have done since the heyday of simple foragers. Many of the nations that sign on to these treaties and conventions do remain quite patriarchal, but such legal mechanisms create a world climate of opinion and often can be used by local activists to leverage more change from reluctant governments and societies.

This chapter has argued that female techno-economic empowerment has been another positive and sweeping trend. Here are some of the dividends (Blumberg, 2005, 2009):

- Studying **international armed conflict**, Caprioli (2000) found an inverse link to female equality: (1) a 5 percent rise in female labor force participation was linked to a state being 495 percent *less* likely to fight its neighbors; (2) a one-third drop in fertility was linked to a state being 467 percent less likely.
- For **internal armed conflict**, she found (2005) that nations with only 10 percent female LFP were ***thirty times (3,000 percent!) more*** likely to have it than those with 40 percent female LFP.
- Greater female involvement in business, the labor force, or parliament was linked to *less* **corruption** (King and Mason, 2001).
- Copeland (2006) found an inverse link in sub-Saharan Africa between women who controlled the spending of their own cash income and the **HIV/AIDS** prevalence rate.
- In my own fieldwork in developing countries, for example in Northern Thailand (Blumberg, 2002), I found that rural women with more economic power have more say in **environmental/land use** decisions at both family and community levels and tend to opt for conservation (after all, in deforested rural areas, where the hydrological cycle has been broken, they are the ones who must go farther for water in the dry season and to collect firewood year round).
- Women are driving the success of the world **microfinance movement** (involving tiny loans to microentrepreneurs): by the end of 2007, 3,552 microfinance institutions (MFIs) reached over 155 *million* clients. At time of first loan, 106.6 million were among the poorest and 83.4 percent of these were women. Their loans affected *c.* 533 million people (Daley-Harris, 2009). Microcredit now impacts whole nations: the vast majority of Bangladesh villages have MFIs, with an overwhelming proportion of their clientele being women; World Bank data indicate that all this has lowered the country's poverty rate. (Mohammed Yunus, founder of the Grameen Bank, the best-known MFI, won the 2006 Nobel Peace Prize for his work. Its clients have shifted from 64 percent

male in the early 1980s to about 97 percent female in recent years, with women clients maintaining a 98 percent repayment rate [Blumberg, 2001b, 2006].) Women not only are more likely to repay loans – and on time – but they are also more likely to channel any profits not reinvested in the business to their children's welfare. "Best practices" MFIs charge market rates of interest and strive to turn a profit, so clients of the most rigorous (those with computerized management information systems [MIS], to track loan repayment) tend to feminize over time.[17]

Increasingly, it appears that whether the trend toward female empowerment and equality in our globalized world continues – or not – can have planetary consequences. If it is constrained or stymied by, say, an upsurge in widespread terrorism/asymmetric warfare, perhaps linked to increased religious fundamentalism, the consequences could be the reversal of most of the positive trends noted in the preceding bullets. Indeed, the impact might be enough to tip the balance concerning use of nuclear weapons (which could produce nuclear winter) or, conversely, efforts to reduce our carbon footprint (which could lead to the intensification of global warming). Similarly, hampering women's ability to produce food[18] or income could hinder efforts to provide macro- and micro-level food security, with resultant decreases in material and psychological well-being and, potentially, increases in national and international instability.

In conclusion, it is too soon to predict the outcome of the race between the accelerating transformations linked to the apparent rise of a new global knowledge/life techno-economic base and the often capricious impact of rising terror, climate change, volatile commodity prices, and other forms of uncertainty. Nonetheless, we *can* say that a continuation of the trend (back) toward the "50–50" line of gender equality is more likely

[17] According to Richard Roberts, FAO/Rome, in a March 1999 interview, a UN–FAO study in the Philippines found that among MFIs with computerized MIS the proportion of women clients rose to 70 percent, versus less than half in MFIs without computerized MIS. In the former, the MIS data showed that women had higher repayment rates. Then, due to the "best practice" of giving loan officers a good percentage of their pay in the form of a bonus – mostly for keeping their portfolios below a target rate of arrears (usually under 5 percent) – they found it in their interest to seek out the best-performing clients. They already could see that these tended to be women and targeted them, resulting in a rising proportion of female borrowers. (To add a final personal note, since 1985 I have worked with and researched microfinance programs in sixteen countries in Latin America, Africa, Asia, and Eastern Europe. I have seen the extraordinary – and gendered – trajectory of MFI growth and impact.)

[18] According to the UN Food and Agriculture Organization (FAO, 2008), "rural women are the main producers of the world's staple crops – rice, wheat, maize – which provide up to 90% of the rural poor's food intake . . . Their contribution to secondary [food] crop production, such as legumes and vegetables, is even greater."

to be linked to the "best of times" – given all the benefits of women's empowerment – while the reverse could help usher in the "worst of times."

References

Baumann, H. (1928). The division of work according to sex in African hoe culture. *Africa: Journal of the International African Institute*, 1(3), 289–319.

BBC News (2008). Women to rule Rwanda parliament. Sept. 20. Retrieved September 20, 2008, from http://newsvote.bbc.co.uk/mpapps/pagetools/print/news.bbc.co.uk/2/hi/africa/7620816.stm.

Birdsell, J. B. (1968). Some predictions for the Pleistocene based on equilibrium systems among recent hunter gatherers. In R. Lee and I. DeVore (Eds.), *Man, the Hunter* (pp. 229–240). Chicago: Aldine.

Blair-Loy, M. (2003). *Competing Devotions: Career and family among women executives.* Cambridge, MA: Harvard University Press.

Blumberg, R. L. (1978). *Stratification: Socioeconomic and sexual inequality.* Dubuque, IA: W. C. Brown.

(1981). *But Are We Helping the People?* Washington, DC: Agency for International Development/Asia Bureau.

(1984). A general theory of gender stratification. *Sociological Theory*, 2, 23–101.

(1986). *Through the Window of Opportunity: Enhancing the World Bank's agricultural development project in Imo State, Nigeria by enhancing extension and incentives for women farmers.* Washington, DC and La Jolla, CA: World Bank and University of California, San Diego.

(1988). Income under female vs. male control. *Journal of Family Issues*, 9, 51–84.

(1989a). *Making the Case for the Gender Variable: Women and the wealth and well-being of nations.* Washington, DC: Agency for International Development, Office of Women in Development (PN-ABC-454).

(1989b). Toward a feminist theory of development. In R. A. Wallace (Ed.), *Feminism and Sociological Theory* (pp. 161–199). Newbury Park, CA: Sage.

(1991). Introduction: The "triple overlap" of gender stratification, economy, and the family. In R. L. Blumberg (Ed.), *Gender, Family, and Economy: The triple overlap* (pp. 7–32). Newbury Park, CA and London: Sage.

(1992). *African Women in Agriculture: Farmers, students, extension agents, chiefs.* Morrilton, AK: Winrock International Institute for Agricultural Development/Development Studies Paper Series.

(1993). Poverty vs. "purse power": The political economy of the mother–child family. In J. Mencher and A. Okongwu (Eds.), *Where Did All the Men Go? Women-headed households in cross-cultural perspective* (pp. 13–52). Boulder, CO: Westview Press.

(1995). Engendering wealth and well-being in an era of economic transformation. In R. Blumberg, C. Rakowski, I. Tinker, and M. Monteon (Eds.), *Engendering Wealth and Well-being: Empowerment for global change* (pp. 1–14). Boulder, CO: Westview Press.

(1998). Gender equality as a human right. *INSTRAW News*, 28, 7–20.

(2001a). Adventures along the gender frontier: Encounters with gender equality in Ecuador and Thailand and glimpses in Guinea-Bissau and China. Paper presented at the International Conference on Gender and Equity Issues: Humanistic Considerations for the 21st Century, January, Bangkok. Also published in the *Conference Proceedings*, Bangkok, Thailand: Srinakharinwirot University Press.

(2001b). We are family: Gender, microenterprise, family work and well-being in Ecuador and the Dominican Republic – with comparative data from Guatemala, Swaziland, and Guinea-Bissau. *History of the Family: An international quarterly*, 6, 271–299.

(2002). A "natural experiment" for gender stratification theory? The Lao of Northeast Thailand and Laos. Paper presented at the meetings of the American Sociological Association, August, Chicago, IL, USA.

(2004a). Extending Lenski's schema to hold up both halves of the sky – a theory-guided way of conceptualizing agrarian societies that illuminates a puzzle about gender stratification. *Sociological Theory*, 22(2), 278–291.

(2004b). Women's rights, land rights and human rights dilemmas in East Africa. *Journal of Development Alternatives and Area Studies*, 23(3–4), 17–32.

(2004c). Climbing the pyramids of power. Alternative routes to women's empowerment and activism. In P. Smith, J. Troutner, and C. Hunefeldt (Eds.), *Promises of Empowerment: Women in East Asia and Latin America* (pp. 60–87). Lanham, MD: Rowman & Littlefield.

(2005). Gender, economy, and kinship in complex households among six U.S. ethnic groups: Who benefits? Whose kin? Who cares? In L. Schwede, R. Blumberg, and A. Chan (Eds.), *Complex Ethnic Households in America* (pp. 248–279). Lanham, MD: Rowman & Littlefield.

(2006). How mother's economic activities and empowerment affect early childhood care and education for boys and girls. Background paper for *Education for All Global Monitoring Report (2007): Strong foundations – early childhood care and education.* Paris: UNESCO 2007/ED/EFA/MRT/PI/6 (electronic version 2007).

(2009). Women's economic empowerment vs. disempowerment: From the "magic potion" for development to the "four horsemen of the Apocalypse"? Paper presented at the UNESCO Women's Studies and Gender Researching Network Conference, February, Bangkok, Thailand.

Blumberg, R. L., Brachet-Marquez, V., Cortes, F., and Rubalcava, R. (1992). Women's "purse power" in the household: Reducing favoritism towards boys' schooling in Santiago, Chile. Paper presented at the meetings of the American Sociological Association, August, Pittsburgh, PA, USA.

Blumberg, R. L., and Coleman, M. T. (1989). A theory-guided look at the gender balance of power in the American couple. *Journal of Family Issues*, 10(2), 225–249.

Blumberg, R. L., and Salazar, A. (2006). Can a focus on survival and health as socio-economic rights help some of the world's most imperiled women in a globalized world? Cases from Ecuador, Ukraine, and Laos. Presented at UNESCO Conference on Gender and Human Rights, July, Nantes, France.

Blumstein, P., and Schwartz, P. (1991). Money and ideology: Their impact on power and the division of household labor. In R. Blumberg (Ed.), *Gender, Family, and Economy: The triple overlap* (pp. 261–288). Newbury Park, CA and London: Sage.

Boserup, E. (1970). *Woman's Role in Economic Development*. New York: St. Martins.

Brown, J. (1975). Iroquois women: An ethnohistoric note. In R. R. Reiter (Ed.), *Toward an Anthropology of Women* (pp. 235–251). New York: Monthly Review Press.

Bryson, J. C. (1981). Women and agriculture in sub-Saharan Africa: Implications for development (an exploratory study). *Journal of Development Studies*, 17(3), 29–46.

Burger, C., Abbot, G., Tobias, S., Koch, J., Vogt, C., Bievenue, L., Carlito, D., Sosa, T., and Strawn, C. (2007). Gender equity in science, engineering and technology. In S. Klein *et al.* (Eds.), *Handbook for Achieving Gender Equity Through Education* (pp. 255–279). Mahwah, NJ: Erlbaum.

Caprioli, M. (2000). Gendered conflict. *Journal of Peace Research*, 37(1), 51–68.

(2005). Primed for violence: The role of gender inequality in predicting internal conflict. *International Studies Quarterly*, 49, 161–178.

Clignet, R. (1970). *Many Wives, Many Powers*. Evanston, IL: Northwestern University Press.

Cohoon, J. McGrath, and Aspray, W. (Eds.) (2006). *Women and Information Technology: Research on underrepresentation*. Cambridge, MA: The MIT Press.

Collins, R. (1971). A conflict theory of sexual stratification. *Social Problems*, 19, 3–20.

(1975). *Conflict Sociology: Toward an explanatory science*. New York: Academic Press.

Collins, R., Chafetz, J., Blumberg, R. L., Coltrane, S., and Turner, J. (1993). Toward an integrated theory of gender stratification. *Sociological Perspectives*, 36, 185–216.

Coltrane, S., and Collins, R. (2001). *Sociology of Marriage and the Family: Gender, love, and property*, 5th edn. Belmont, CA: Wadsworth.

Copeland, C. (2006). Social determinants of high HIV prevalence in sub-Saharan Africa. Honors thesis, University of Virginia.

Cowan, R. S. (1983). *More Work for Mother: The ironies of household technology from the open hearth to the microwave*. New York: Basic Books.

Daley-Harris, S. (2009). *State of the Microcredit Summit Campaign Report 2009*. Washington, DC: The Microcredit Summit Campaign. Also at: www.microcreditsummit.org/SOCR/SOCR2009_English.pdf.

De Waal, F. (2005). *Our Inner Ape*. New York: Riverhead Books.

Diamond, J. (1997). *Guns, Germs, and Steel: The fate of human societies*. New York: Norton.

Economist. (2006a). The importance of sex, 378(8473), April 15, 16.

(2006b). A guide to womenomics, 378(8473), April 15, 73–74.

Eloundou-Enyegue, P. M., and Calves, A. E. (2006). Till marriage do us part: Education and remittances from married women in Africa. *Comparative Education Review*, 50(1), 1–20.

Engelman, Robert. (2008). *More: Population, nature and what women want.* Washington, DC: Island Press.

Faiola, A. (2008). Women rise in Rwanda's economic revival. *Washington Post,* May 16, A1, A12.

FAO. (2008). *Gender and Food Security: Agriculture.* Rome: FAO (Food and Agriculture Organization). Retrieved June 1, 2009 from www.fao.org/gender/en/agri-e.htm.

Fenstermaker, S., West, C., and Zimmerman, D. H. (1991). Gender inequality: New conceptual terrain. In R. Blumberg (Ed.), *Gender, Family and Economy: The triple overlap* (pp. 289–307). Newbury Park, CA and London: Sage.

Fisher, K., Egerton, M., Gershuny, J., and Robinson, J. (2006). Gender convergence in the American Heritage Time Use Study (AHTUS). Social Indicators Research. DOI 10.1007/s11205-006-9017-y.

Frisch, R. (1978). Nutrition, fatness, and fertility: The effect of food intake and reproductive ability. In W. H. Mosley (Ed.), *Nutrition and Human Reproduction* (pp. 91–122). New York: Plenum Press.

Fry, D. P. (2006). *The Human Potential for Peace: An anthropological challenge to assumptions about war and violence.* New York and Oxford: Oxford University Press.

Geertz, C. (1963). *Agricultural Involution.* Berkeley, CA: University of California Press.

Goldin, C. (1986). The female labor force and American economic growth. In S. Engerman and R. Gallman (Eds.), *Long-term Factors in American Economic Growth* (pp. 557–604). Chicago: University of Chicago Press.

Gonsoulin, M. (2005). Women's rights and women's rites: Religion as the embodiment of kinship and gender ideals. Unpublished doctoral dissertation, University of Virginia.

Gough, K. (1971). The origin of the family. *Journal of Marriage and the Family,* 33(4), 750–771.

Haraway, D. (1985). A manifesto for cyborgs: Science, technology, and socialist feminism in the 1980s. *Socialist Review,* 15(2), 65–107.

(1991). *Simians, Cyborgs, and Women: The reinvention of nature.* New York: Routledge.

Harriss-White, B. (1999). Gender-cleansing. In R. S. Rajan (Ed.), *Signposts: Gender issues in post Independence India* (pp. 124–153). New Delhi, India: Kali for Women.

Henderson, M. (2007). Women may have invented weapons. *New York Times,* February 24, 15.

Hubbard, R. (2001). Science, facts, and feminism. In M. Wyer, M. Barbercheck, D. Geisman, H. O. Ozturk, and M. Wayne (Eds.), *Women, Science and Technology: A reader in feminist science studies* (pp. 153–160). New York: Routledge.

IPU (Inter-Parliamentary Union). (2008). *Women in Parliament: Situation as of 31 July 2008.* Retrieved September 20, 2008, from www.ipu.org/wmn-e/arc/classif310708.htm.

ITAA (Information Technology Association of America). (2005). *Building the 21st Century Information Workforce: Groups underrepresented in the IT workforce,*

Task Force Report. Retrieved October 4, 2007, from web.archive.org/web/20041229120144/http://www.itaa.org/workforce/studies/recruit.htm.

Kelly, R. L. (1995). *The Foraging Spectrum: Diversity in hunter-gatherer lifeways.* Washington, DC: Smithsonian Institution Press.

King, E. M., and Mason, A. D. (2001). *Engendering Development: Through gender equality in rights, resources, and voice.* Washington, DC and New York: The World Bank and Oxford University Press.

Kishor, S. (1993). May God give sons to all: Gender and child mortality in India. *American Sociological Review*, 58(2), 247–265.

Kolata, G. B. (1974). !Kung hunter-gatherers: Feminism, diet and birth control. *Science*, 185, 932–934.

LaFee, S. (2007). Mothers of invention: While the men were out hunting, Stone Age women created a future. *San Diego Union-Tribune*, April 26, E1, E8.

Lee, R. B. (1968). What hunters do for a living, or how to make out on scarce resources. In R. B. Lee and I. DeVore (Eds.), *Man, the Hunter* (pp. 30–49). Chicago: Aldine.

(1969). !Kung Bushmen subsistence: An input–output analysis. In A. P. Vayda (Ed.), *Environment and Cultural Behavior* (pp. 47–79). Garden City, NY: Natural History Press.

(1984). *The Dobe !Kung.* Fort Worth, TX: Holt, Rinehart, and Winston.

Lee, R. B., and DeVore, I. (Eds.) (1968). *Man, the Hunter.* Chicago: Aldine.

Lenski, G. E. (1966). *Power and Privilege.* New York: McGraw-Hill.

(2005). *Ecological-evolutionary Theory: Principles and applications.* Boulder, CO: Paradigm.

Levinson, D. (1989). *Violence in Cross-cultural Perspective.* Newbury Park, CA: Sage.

Marshall, L. (1965). The !Kung Bushmen of the Kalahari Desert. In J. Gibbs (Ed.), *Peoples of Africa* (pp. 243–278). New York: Holt.

Melymuka, K. (2001). If girls don't get IT, IT won't get girls. *Computer World*, January 8, 35(2), 44.

Murdock, G. P. (1967). Ethnographic atlas: A summary. *Ethnology*, 6, 109–236.

Murdock, G. P., and Provost, C. (1973). Factors in the division of labor by sex: A cross-cultural analysis. *Ethnology*, 8, 329–369.

NCES (National Center for Education Statistics). (2002). *Table 282: Earned degrees in computer and information sciences conferred by degree-granting institutions, by level of degree and sex of student, 1970–71 through 2000–2001.* Higher Education General Information Survey and Integrated Post-secondary Education Data System. Retrieved October 4, 2007, from http://en.wikipedia.org/wiki/Women%2C_girls_and_information_technology.

Nolan, P., and Lenski, G. (2009). *Human Societies: An introduction to macrosociology*, 11th edn. Boulder, CO: Paradigm Publishers.

NSF (National Science Foundation). (2000). *Land of Plenty: Diversity as America's competitive edge in science, engineering and technology*, cawmset0409. Arlington, VA: NSF. Cited in Burger *et al.*, 2007 (pp. 255–279).

(2004). *Women, Minorities and Persons with Disabilities in Science and Engineering.* Arlington, VA: NSF. Cited in Burger *et al.*, 2007 (pp. 255–279).

(2006). *First Time S&E Graduate Enrollment of Foreign Students Drops for the Third Straight Year.* Arlington, VA: NSF. Retrieved June 1, 2009 from www.nsf.gov/statistics/infbrief/nsf06321. Cited in Burger *et al.*, 2007 (pp. 255–279).
Oppenheimer, V. K. (1973). Demographic influence on female employment and the status of women. *American Journal of Sociology*, 78, 946–961.
Sahlins, M. D. (1968). Notes on the original affluent society. In R. Lee and I. DeVore (Eds.), *Man, the Hunter* (pp. 85–89). Chicago: Aldine.
Saito, K., and Weidemann, J. (1990). *Agricultural Extension for Women Farmers in Africa.* Washington, DC: The World Bank/Women in Development Working Papers.
Slocum, S. (1975). Woman the gatherer: Male bias in anthropology. In R. R. Reiter (Ed.), *Toward an Anthropology of Women* (pp. 36–50). New York: Monthly Review Press.
Sudarkasa, N. (1973). *Where Women Work: A study of Yoruba women in the marketplace and in the home.* Anthropological Papers, Museum of Anthropology, No. 53. Ann Arbor, MI: University of Michigan.
Sullivan, O., and Coltrane, S. (2008). Men's changing contribution to housework and childcare. Paper presented at the 11th annual conference of the Council on Contemporary Families, April, Chicago, IL, USA.
Tang, J. (2006). *Scientific Pioneers: Women succeeding in science.* Lanham, MD: University Press of America.
Toffler, A. (1990). *Powershift.* New York: Bantam Books.
Turnbull, C. M. (1961). *The Forest People: A study of the Pygmies of the Congo.* New York: Simon & Schuster/Touchstone Books.
(1981). Mbuti womanhood. In F. Dahlberg (Ed.), *Woman the Gatherer* (pp. 205–219). New Haven, CT: Yale University Press.
United Nations. (1987). *Fertility Behaviors in the Context of Development.* New York: United Nations, Department of International Economic and Social Affairs, ST/ESA/SER.A/100.
Wajcman, J. (1991). *Feminism Confronts Technology.* Cambridge, MA: Polity Press.
(2000). Reflections on gender and technology studies: In what state is the art? *Social Studies in Science*, 30(3), 447–464.
Wikipedia. (2007). *Women, Girls and Information Technology.* Retrieved October 4, 2007, from http://en.wikipedia.org/wiki/Women,_girls_and_information_ technology.
Wyer, M., Barbercheck, M., Geisman, D., Ozturk, H. O., and Wayne, M. (Eds.) (2001). *Women, Science, and Technology: A reader in feminist science studies.* New York: Routledge.

10 Technology and well-being: designing the future

Yair Amichai-Hamburger

This book has demonstrated that technology has enhanced many aspects of our lives, but its effects are complex. This final chapter will discuss how we can increase the positive impact of technology and minimize some of its more dangerous aspects. In the first chapter, Rodman and Fry utilized the symbols of Yin and Yang to demonstrate the contradictory effects technology can have on our well-being. They pointed out, for example, that while portable music players such as the iPod provide great pleasure to their users, they may also isolate listeners by allowing them to tune out the world around them. Amichai-Hamburger and Barak (in Chapter 2) suggested that the Internet can provide an excellent psychological environment enabling people to find support for many kinds of difficulties they may be facing. However, this same psychological environment may create addictive behavior among its users.

This chapter will begin by assessing the major roles played by technology within our society. It will ask whether technology has the answers to many of the current and future challenges facing society or whether it is a tool that may be harnessed to achieve our real goals. In other words, is technology a means to an end or does it constitute the end? This chapter will initially explore some of the more worrying aspects of our highly technological society, and then go on to discuss ways in which present technological achievements and those of the future may be used to promote the well-being of the societies in which they operate. We see technology neither as the fire-breathing dragon, placing our lives in imminent peril (as it is perceived by the Neo-Luddism movement, among others), nor as the knight on a white charger coming to save the world from all its ills (as the techno-Utopians would have us believe). This chapter will investigate in greater depth the overall interaction between technology and well-being in both its positive and negative aspects.

The current role of technology in our lives

Covey (1989) related a fascinating personal story by Frank Koch that shows the importance of clarifying which are the real values that individuals wish to live by ("principles" as defined by Covey), as opposed to those that society attempts to impose on us.

> Two battleships assigned to the training squadron had been at sea on maneuvers in heavy weather, for several days. I was serving on the lead battleship and was on watch on the bridge as night fell. The visibility was poor with patchy fog, so the captain remained on the bridge keeping an eye on all activities.
>
> Shortly after dark, the lookout on the wing of the bridge reported, "Light, bearing on the starboard bow."
>
> "Is it steady or moving astern?" the captain called out.
>
> Lookout replied, "Steady, captain," which meant we were on a dangerous collision course with that ship.
>
> The captain then called to the signalman, "Signal that ship: We are on a collision course, advise you change course 20 degrees."
>
> Back came a signal, "Advisable for you to change course 20 degrees."
>
> The captain said, "Send I'm a captain, change course 20 degrees."
>
> "I'm a seaman second class," came the reply. "YOU had better change course 20 degrees."
>
> By that time the captain was furious. He spat out, "Send, I'm a battleship. Change course 20 degrees."
>
> Back came the flashing light, "I'm a lighthouse."
>
> We changed course. (Covey, 1989: 33)

According to Covey, the lighthouse in this story can be seen as representing unchanging human values that are relevant across time and throughout the world, and do not change according to which captain is at the helm. This story may also help us to understand the role that should, in our view, be allotted to technology in modern life and may also help to answer the question of how we can increase the positive impact of technology on our well-being. In the discussion below the components of the lighthouse story are employed to clarify major aspects of the relationship between technology and well-being. Topics include: (1) leadership or lack of it; (2) our society's values; and (3) colliding with the rocks.

Leadership or lack of it

Leadership, in contrast to management, is about having a vision and leading the people toward its fulfillment. Management without leadership has been likened to the arranging of the chairs on the Titanic. It is the leader who is capable of noticing the iceberg, avoiding a collision, and steering the ship in the right direction. Yukl (2002) suggested that transformational leaders have a vision that is germane to their followers, build strategies to bring this vision to life, and then build the confidence in their followers that this vision can be achieved, despite the fact that there are major challenges ahead.

Today, much of the media use technology to spread messages that are not positive or beneficial to society. In some cases, modern political leaders appear prepared to take on the negative undue influence of media technology; however, on deeper examination, it seems that in many cases their actions do not tally with their words (Hamburger, 2000). In some cases, this lack of action in relation to technology stems from mistaken priorities. It may also be the result of the influence of the interests that support the particular leader. When a leader or a party is allied to large business or media concerns, his/her ability or willingness to exert authority or control over these interests is significantly reduced.

Below we analyze some of the major values in today's Western society which, at best, do not promote and, at worst, clash with those permanent unchanging values symbolized by the lighthouse.

Our society's values

For many living in a capitalist society, it seems that its main values are: individuality, mass consumption, and the time/money equation.

Individuality In many cases, such a society allows greater freedom than other societies. It gives the individual what seems to be a wider range of options. However, some of the roads to individuality in our society do not really take us there. Many of the advertisements around us suggest that we can fulfill ourselves as individuals through the acquisition of gadgets. For example, we are told that the purchase of *this* new cell phone will instantly transform us into cool, significant individuals. It is a naïve and almost paradoxical assumption that the purchase of *mass produced* technology can help us each to become significant as *individuals*. The illusion of individuality is demonstrated particularly well in a short scene in the 1979 film *Life of Brian*. In the movie, Brian has inadvertently accumulated a large crowd of disciples who follow him, believing he is the Messiah. In this extract, Brian is trying to rid them of this misconception (emphasis is mine):

BRIAN: Please, please, please listen! I've got one or two things to say.
THE CROWD: Tell us! Tell us both of them!
BRIAN: Look, you've got it all wrong! You don't NEED to follow ME; you don't NEED to follow ANYBODY! You've got to think for yourselves! You're ALL individuals!
THE CROWD: Yes! We're all individuals!
BRIAN: You're all different!
THE CROWD: Yes, we ARE all different!
MAN IN CROWD: I'm not...
THE CROWD: Ssh!

To a great extent, individuality in the Western world is an illusion, as this story reflects. What appears to be individuality is, in many cases, just the result of a campaign of persuasion run by large empires with the aim of convincing us that we will become significant individuals if we buy their products. Such processes may be conveyed through the deployment of cultural heroes, for example movie stars, sportsmen, and models. Through a process of identification, people are induced to buy commodities (allegedly) used or owned by a celebrity. Carl Rogers (1961) warned against the cultural brainwashing found in modern society, which, he believed, would influence the way in which young people defined their ideal self. Adapting an ideal self to the mores of modern society is likely to prevent the individual from being able to express his or her real individuality, and this may lead to psychological complications.

Mass consumption Media channels encourage people in the modern world to consume all the time: "Shop till you drop." This has created a very materialistic approach to life. Another reason for rampant consumerism has been suggested by Ben-Shahar (2007). He argued that society's excessive belief in technology has led to materialistic thinking. Easterlin (2003) suggested that most people believe that a 20 percent raise in their salary would make them happy; however, when they achieve it, it does not affect their feelings of well-being. Research has shown that those people who placed a high value on material goals were found to be less happy than those who were less materialistic (Kasser, 2002). Materialism was related to lower self-esteem, more narcissism, more social comparison, less empathy, and more conflict in relationships (Kasser *et al.*, 2004).

Time/money equation "Time equals money" is a slogan constantly heard in modern society. It refers to the need for total efficiency, and it is likely to lead to a focus on the here and now, at the expense of long-term goals and a recognition of the process that leads to the result. As an industrial consultant, I was once asked to analyze a high-tech company. After a careful assessment, I suggested that senior management should discontinue their policy of promoting an artificial atmosphere of a constant crisis in their organization; while that policy may have provided a good excuse to keep the workforce constantly on their toes and squeeze more work out of them, it simultaneously created significant burnout and a reduction in creativity. Perhaps unsurprisingly, I was never invited back.

It seems that we, as a society, have become used to functioning under pressure, and this is now perceived as part of a normal life. However, stress clearly has a negative effect on different aspects of our lives, from our personal well-being to our functioning at work (Bakker *et al.*, 2000). As Amabile *et al.* (2002) put it, when creativity is under the gun, it will usually end up getting killed. The lack of clear boundaries between work and home is also a major cause of stress in our society (Harrison *et al.*, 2004). As a friend of mine, a human-resource manager in a high-tech company, pointed out: "They gave me a mobile phone as a present so they can own me 24 hours a day. They gave me a portable computer, so my office is now with me all the time and I cannot block myself from work pressure; it is with me all the time 24 hours a day." It seems that every organizational present has a price and that price is unremitting pressure. The effects of the lack of separation between work and home may be so great that, for some people, the only way to reenergize is to go away on vacation. However, as Fisher (2005) related, 34 percent of people in the US reported being in touch with the office even during vacations. It should come as no surprise that many such people return to work as stressed as when they left, and in some cases even more so. It seems that modern society is in danger of replacing quality of life with standard of living. We are under continuous pressure to achieve and, without noticing it, this may come at the expense of the most important components of our lives.

When communication technology is utilized to promote the kind of values described above, it is very hard to resist its influence.

As early as 1970, Ellul warned that technology is no longer a means to achieve human objectives, but rather has become the master controlling humans and their environment (Ellul, 1970). A direct demonstration of this influence may be seen in the power of media to run campaigns of persuasion, for example to enhance sales (Cialdini, 1984). Harrington (1966) warned of the scientific methods utilized by the advertising industry in order to galvanize public opinion toward private causes. These seek to direct the free will and freedom of choice of the consumer so as to bring about maximum profit. Since we are surrounded by the media, each of their major campaigns has the potential to brainwash us. Klein (2000) pointed out that, in modern life, both politicians and commodities are sold by linking them with human values. This call to our values from different types of media makes such campaigns even more difficult to resist. As Diener and Seligman (2004) explained, our freedom to choose is determined by the leading industrial companies; this is, in fact, the freedom to choose between a limited number of given options.

Postman (1985), analyzing a potential futurist dictatorship, pointed out that while it was the book confiscators who frightened Orwell (1949), for Huxley (1932) the real fear was the day after, when people no longer wish to read books. In Orwell's world, our freedom is taken by force; in Huxley's, it is removed without our even noticing. The public has yet to appreciate, explained Postman (1985), that technology is an ideology. A clear example of the type of danger that Postman warned against may be seen in the sophistication of modern computer technology which has the capacity to learn the psychological profile of its users. This knowledge can help companies manipulate us with even greater ease. For example, the management of Google announced recently that it has the technology to build an exceptionally sophisticated personal profile of individual users. According to Google, their techniques allow them to learn more about an individual than he/she knows about him/herself (Amichai-Hamburger, 2007). This power clearly carries with it many features that may endanger human freedoms.

Below we describe what we see as a major threat to society: that in a situation in which there is a lack of genuine leadership and a fading of the real "lighthouse" values, we are in serious danger of colliding with the rocks.

Colliding with the rocks

Goleman (1995) suggested that the world is coming out of the major depression of the twentieth century, characterized by the "Age of Anxiety," and moving toward the "Age of Melancholy." Diener and Seligman (2004) cited strong evidence to suggest that young people today experience more depression than those of previous generations. They reported that, since the beginning of the 1970s, the frequency of anxiety and depression has increased fourfold, and pointed out that, as every objective index of well-being, for example education, availability of food, and power of buying, has risen, all indices of mental well-being have demonstrated a consistent decline (Seligman, 2002). Strenger (2004) suggested that, in modern urban society, young people have no authority figures; they do not even have to rebel against their parents because their parents are not relevant any more. Postman (1982) argued that television, with its adult content, has shortened childhood; TV messages are flat, empty, focus on the here and now, and do not demand any mental effort from the child.

This may be seen as part of a long-term trend in our society toward the weakening of social contacts (see, e.g., Durkheim, 1893; Weber, 1958). Putnam (2000) pointed out that social capital, a term he defines as "relating to the connections among individuals – social networks and the norms of reciprocity and trustworthiness that arise from them" (2000: 19), is presently in a state of serious decline in America. He argued that the social fabric of American society is dissolving, social clubs and organizations are losing their centrality, and increasing numbers of people live isolated lives. Twenge (2002) suggested that rising dysphoria in the US is partly due to the breakdown of social connectedness.

In other words, returning to our metaphor, it seems that our ship, our society, has a captain with no clear vision of how to lead the ship toward a safe shore – that is, how to enhance well-being in our society – and we are coming closer to a major collision with the rocks – that is, a collapse of the healthier parts of society.

Clearly, society is dependent on technology, and technology can and does enhance our lives in many ways. However, as we pointed out previously, technology does not intrinsically have values; without sufficient ethical guidance, technology may harm our well-being. Below we describe our suggestions for resolving this situation.

Toward a solution

One of the key aims of a democratic government is to promote the good life: A flourishing society where citizens are happy, healthy, capable, and engaged, in other words with high levels of well-being (Marks and Shah, 2005). Below we assess the main components necessary to promote the positive impact of technology on our well-being. We will discuss: (1) recognizing the lighthouse; (2) the captain or the leadership story; and (3) utilizing the lighthouse.

Recognizing the lighthouse

If we want to become a society that shapes its own future, we have to recognize our lighthouse, that is, recognize the permanent values which are necessary to protect us and prevent our hazardous collision with the rocks. When looking at these values in terms of what makes us happy, there is a clear attraction toward economic indicators as a gauge of well-being: first, because they are easily measurable and, second, because traditional thinking informs us that wealthy people are happier than those who are not. However, studies have shown that, once a certain level of basic needs has been met, economic prosperity is not an accurate measure of well-being and, therefore, other parameters must be sought (Diener and Seligman, 2004; Marks and Shah, 2005). According to Seligman (2002), people wish to feel that they have achieved their well-being and that it is not just a matter of luck. They are not seeking to live a life composed of short stimuli that lead to short-term feelings of happiness, but, rather, they are looking for a long-term feeling of well-being. Below we discuss the bases for long-term well-being, which we see as being synonymous with the unchanging "lighthouse values."

Layard (2005) argued that seven components are central to happiness:

- Family relationships – being able to give one another love and comfort.
- Finances – especially when people are living on the margins of poverty.
- Work – provides not only income, but may also give an extra meaning to life. When people have a high degree of control over what they do, work is fulfilling.
- Community and friends – communities that produce a sense of trust and belonging have a strong impact on our happiness.
- Health – we can adapt to many things, but chronic illness is very difficult to cope with.
- Personal freedom – there is evidence that people living in stable and peaceful societies in which they have a voice and an ability to follow their interests and in which institutions are accountable are happier.
- Personal values – people are happier if they are able to appreciate what they have, whatever it is; if they do not compare themselves with others; and if they school their own moods.

Diener and Seligman (2002) studied very happy people. They compared happy people to those who are less happy. The only external factor that distinguished the two groups was the presence of rich and satisfying social relationships. In other words, spending meaningful time with friends, family, and romantic partners was necessary (though not in itself sufficient) for happiness. Diener and Seligman argued that marriage and long-term cohabiting relationships have beneficial effects on well-being. Having family networks and intimate friendships is also an important contributor, as is belonging to a community and a social group. Putnam (2000) suggested that community engagement not only improves the well-being of those involved, but also improves the well-being of others. The relationship is positive in both directions: Involvement increases well-being and happy people tend to be more involved in their communities. Therefore, interventions in this area should lead to a positive upward spiral (see, also, Helliwell and Putnam, 2004).

Deci and Ryan (1985) studied the conditions that foster positive human potential. They identified three significant internal needs: *autonomy* – feeling that your activities are self-chosen and self-endorsed; *competence* – feeling that you are effective in your activities; and *relatedness* – feeling a sense of closeness with others. They went on to develop the self-determination theory (SDT), which hypothesized that these three needs are crucial for behavioral self-regulation, personality development, and positive well-being (Ryan *et al.*, 1997). Interestingly, Ryan and Deci (2007) argued that competence, autonomy, and relatedness serve as a platform for the enhancement of intrinsic motivation in a child, which is essential for realizing the positive potential of humankind.

Sheldon *et al.* (2001) posed the question: What are the most fundamental needs for humans? In their findings, they confirmed the centrality of autonomy, competence, and relatedness needs as suggested by the self-determination theory (Ryan and Deci, 2000), and added another component: self-esteem. Sheldon and Elliot (1999) advise people who are seeking greater well-being to focus on the pursuit of: (1) goals that involve growth, connection, and contribution to others, rather than goals which involve beauty, money, or popularity; (2) goals that are relevant and important to oneself, rather than those that are forced upon one. The pursuit of these self-chosen goals is likely to enhance individual well-being in relation to technology.

When assessing the list of components that encourage well-being, we tend toward the self-determination theory, advocating competence, autonomy, and relatedness. Among these factors, relatedness is, we suggest, the one that stands out as a determining factor at many different levels, including those of interpersonal relationships: for example marriage, friendship, and social groups. We suggest that we must aim to harness each one of these basic human needs as part of our permanent value system, "our lighthouse." These needs provide a measure against which to determine whether or not a technological invention is being exploited in a positive way, that is, one that will enhance our well-being.[1]

As we described earlier, one of the problems of technology is that it can be used as a manipulative tool. It appears that developing the skills of autonomy and competence serves as an important basis for promoting an independent mind. However, we suggest that this in itself will not suffice;

[1] I received the endorsement of Deci and Ryan for my transforming their "needs" into "values."

there is another component that is required: that of critical thinking. In other words, we believe that it is necessary to extend people's capacity to develop sound criteria and standards for analyzing and assessing their own thinking, and routinely use those criteria and standards to improve its quality (Elder and Paul, 1994). Another definition of critical thinking relates to "a conscious and deliberate process which is used to interpret or evaluate information and experiences with a set of reflective attitudes and abilities that guide thoughtful beliefs and actions" (Mertes, 1991: 24). Potter (2001) argued that, in order for people to develop the capacity for thinking critically, it is necessary for them to develop cognitive, emotional, aesthetic, and moral skills. Once they are able to think critically, people will be better able to analyze the messages with which they are being bombarded by the media and, where necessary, to reject them. This is clearly not an easy task; however, it is important that individuals develop such skills in order to maintain their freedom and integrity.

To be effective, this process of directing technology toward enhancing well-being requires effective, responsible guidance. Below we discuss the implications of this leadership factor.

The captain or the leadership story

Our society requires leadership that is determined to enhance the positive link between technology and well-being. In our view, this type of leadership would work most effectively as an independent body awarded the task of advising the government on issues regarding technology and well-being. Forty years ago, Toffler (1970) argued for the need to determine the social implications of technological inventions. To this end, he called for the establishment of councils of experts to serve as "technological ombudsmen." This group of ombudsmen would include psychologists, sociologists, economists, and experts of political studies. They would be mandated to advise on the development of technologies that have a high potential to improve people's psychological well-being. Should they conclude that a certain invention has dangerous implications, the council would advise as to its future development, thus limiting the invention's negative potential or, in some cases, regulating its progress or expansion. In many ways, this may be seen as akin to the work of medical ethics boards which consider the moral implications of such issues as human cloning and multiple egg transplants. Postman (1993) expanded Toffler's (1970) idea of technological ombudsmen and called for the employment of professional futurists. It is our view that such a committee should include technological innovators themselves, along with members of the public and representatives of policymakers.

Providing capable leadership and defining a real lighthouse is key to preventing a collision with the rocks and bringing the ship, our society, to a safe shore.

Utilizing the lighthouse

Below, we take three major areas of human life and assess how the values incorporated in the lighthouse may be implemented within them. First, we discuss children and technology; second, technology in the workplace; and, third, leisure time use of technology.

Children and technology It is tremendously important that everyone in society has access to technology. What is crucial, in the case of children, is to ensure that they are exposed to its positive, rather than its negative, aspects. One of the aims of the education system should be to teach children how to utilize technologies, such as the Internet, to empower themselves (Amichai-Hamburger *et al.*, 2008).

From almost its earliest stages, the education of children must include equal access to technology, irrespective of socioeconomic background or gender. One venture which is working toward this end is the One Laptop per Child (OLPC) project (http://laptop.org). This aims to provide children in poorer countries with the opportunity to receive a significant education. In wealthier countries, this type of work tends to be carried out on a more local scale.

Another major challenge in terms of equal access is that of equality of gender. To achieve this, it is necessary to eliminate the gender gap in technological education which presently exists in favor of male students. There are several reasons for this gender differential; included among them are use of differential feedback to girls and boys by teachers (Sadker and Sadker, 1984), a lack of material that relates to girls, and the lack of encouragement that girls receive: girls need to be able to learn by using examples that relate to their own experience (Belensky *et al.*, 1989). Another reason for this gap is that girls have internalized negative stereotypes about learning science (Lantz, 1985). Dealing with these challenges will help to spread the positive aspects of technology in our society.

To return to a point made earlier, it is necessary to help children achieve a sense of autonomy, competence, and relatedness, so that they will go on to develop intrinsic motivation (Ryan and Deci, 2007). Children who are taught this and are encouraged to develop critical minds, and, as adults, are likely to be more capable of relating to technology in healthy and positive ways, including, for example, having the ability to recognize and reject negative persuasion campaigns and avoid Internet addiction.

One environment where children are likely to experience feelings of autonomy and competence is that of Internet games. The game is also a very important means through which experience is formulated (Erikson, 1968). The phenomenon of Internet games is a growing one. Turkle (1995) argued that involvement in Internet identity games is similar to participation in psychodrama. Turkle believes that the Internet supplies an individual with space, warmth, safety, and understanding. This is, in fact, similar to the setting provided by psychotherapy, so that both the Internet and the psychotherapy room may create safe environments in which to rework elements from the past and try out different alternatives for the present and the future. Amichai-Hamburger (2005) argued that the main problem with Turkle's (1995) argument is that the Internet may empower people, but the direction of the empowerment depends on the tendencies of all the people taking part online. Interaction with others on the net has the potential to be supportive and help people to explore their identities if all surfers taking part in the interaction are pro-socially oriented and are capable of giving support. However, this is not always the case; interaction over the net, be it in a chat room or a fantasy community, is equally likely to be unhelpful and in some cases may even be hostile. We suggest the creation of a fantasy game environment where people are rewarded according to the level of help and support they give others. Building such an environment will enhance the likelihood of the Internet being utilized as a tool for helping people. Having the combination of the protected environment and the supportive one will enhance the benefits of the Internet for our well-being.

Relatedness can be enhanced through balanced and directed use of communication technology. Children should be shown how to use the Internet to strengthen ties with friends and family. This is especially relevant when friends or family members are not close by and face-to-face meetings are infrequent. This may have particular relevance for older relatives, since their ability to travel may be limited due to reduced physical health and they are therefore prone to loneliness. Keeping in touch through the Internet may be a way of partially overcoming geographical and physical limitations, and can enhance the relatedness factor for both sides.

The workplace Work has traditionally been perceived as a necessity. However, modern thinking suggests that work should increase well-being (Diener and Seligman, 2004), and should help the worker achieve and express competence, autonomy, and relatedness. Relatedness in a work environment refers to the degree of support the job gives the worker and to the level of connectivity the worker has with his or her fellow workers. It is highly recommended that organizations invest in their workers' well-being, not only for the sake of the individual worker, but for the benefit of the whole organization. People with high levels of well-being perform better at work, are very high on scales of organizational citizenship behavior, and are likely to become an asset to their organizations (Allen and Rush, 1998; Konovsky and Pugh, 1994). Technology in the workplace should be implemented in ways that respect the workers' needs for autonomy, competence, and relatedness. This means, for example, minimizing, as far as possible, the invasion of work into other areas of life. Technology, such as mobile computers and phones, should not be used during workers' private time. One important indicator in job evaluation should be the levels of autonomy, competence, and relatedness it provides.

Autonomy is an important factor in the Job Characteristics Model which argues that jobs should be redesigned to enable people to gain more enjoyment from their work (Fried and Ferris, 1987; Hackman and Oldham, 1980). According to the model, in order to achieve a positive impact, it is necessary that the job require: (1) a variety of skills (that the employee use a variety of skills and talents); (2) task identity (the opportunity to complete an entire piece of work); (3) task significance (the opportunity to have an impact on others); (4) autonomy (employees should have the freedom to plan, schedule, and perform in their tasks as they so wish); and (5) feedback (workers should receive comments as to their effectiveness). While autonomy is explicitly mentioned in this model, competence, too, can be found in terms of use of skills, task identity, task significance, and feedback.

In global organizations, the Internet allows managers to relate directly to their employees, cutting out the intermediate managerial layers in the organization. This enables managers to become significant e-leaders who can approach their workers directly with vision, messages, and feedback on their work (Coovert and Burke, 2005). This is very likely to have a positive impact on workers' well-being.

When organizations implement technology in ways that take the human factor into account, they should aim to create a balance between the technological challenge and the capabilities of the workers. Creating such a balance is likely to give workers the experience of *flow*. This is an optimal experience that stems from people's perceptions of challenges and skills in given situations (Csikszentmihalyi, 1975: 50). Experiencing flow enhances well-being (Moneta and Csikszentmihalyi, 1996: 277).

It is also important that technology does not eliminate the human touch especially in the service-based professions. Postman (1993) warned that the introduction of the stethoscope marked the start of the distance that has developed between patient and doctor. Postman was clearly not rejecting the use of technology in medicine, but was emphasizing that it should not eliminate the interpersonal relationship between patient and doctor. Today's doctors have been compared to garage mechanics, since they look at each body part as a separate identity and do not holistically look at the whole person. In some organizations, clients have become a number on the computer and have lost their human characteristics. Interestingly, when service providers do not relate to an individual on a human level, this may affect not only the client, but also the well-being of the service provider. This is because such a limited approach to the client is likely to harm the feelings of autonomy, competence, and relatedness of the service provider.

Leisure time How we spend our free time serves as a test to demonstrate our true beliefs. Autonomy is achieved by our having the freedom to choose our activities in our leisure time. Competence can be achieved by choosing rewarding leisure activities that give us the feeling that we are being effective. Relatedness can be achieved by spending a substantial part of these activities building and strengthening our relationships with significant others.

However, in today's technological age, even leisure time has become more complex: For example, advertisements for cell phones can create the impression that it is no longer necessary to hug your children; instead, all you need is five minutes' quality time through the little screen (seal you with a kiss). This illusion is expanded through other technological inventions that seemingly sell the idea that people no longer require contact with the outside world; for example, there may be people who live under the false impression that they can create and maintain their intimate relationships exclusively through the net. However, this is not the case; even when a relationship started on the net, there is a human need to feel, touch, and be there (Ben-Ze'ev, 2005). The social relationships we maintain through the net do not always reflect the real quality of those friendships; having four hundred "friends" listed on Facebook does not necessarily reflect four hundred close friendships. The Internet may easily create an illusion of relatedness; however, it is only when we invest in those relationships that our social ties become significant. Technology can obviously contribute a great deal to our maintaining relationships in an efficient way and, although it might not be the first option, innovations such as Facebook are an efficient method of informing your social circle of important events in your life. This does not mean that Facebook will become your main channel of communication, but an important one in addition to the more traditional ones.

Another danger of "technological leisure time," related to the issue of critical thinking, is that of addiction to computer games and Internet games (see Chapter 2). It is our belief that a promotion, both within individuals and within society as a whole, of the components of autonomy, competence, and relatedness, together with an understanding of critical thinking, will strengthen people psychologically and ensure that they are better equipped to resist succumbing to addiction.

Another important value that can promote positive utilization of our free time is an involvement in groups both on- and offline which work for causes to which we relate; for example, ecological or social issues. In addition to being part of a group that strives to help others, our participation has many positive implications for strengthening our social ties and reinforcing a strong sense of autonomy and competence. In many cases, it may also give us opportunities to exploit technology in positive ways.

These are just some examples of how we can implement "the lighthouse" values as they relate to technology and these can be expanded upon to include many other areas of our lives.

Last word

Our well-being has implications for every aspect of our society. It promotes sociability, generosity, creativity, activity, tolerance, health, altruism, economic productivity, and long life (Lyubomirsky *et al.*, 2005). Therefore, promoting the well-being of citizens, both psychologically and physically, should be a major task for governments.

While technology is tremendously important in modern society, it does not have any inherent value system, and it is therefore of pivotal importance that technological development and use are regulated by a set of ethical standards. These standards should be taken from the unchanging values of human society that we have termed "lighthouse values" and should be based on those factors known to promote the well-being of people.

The values we advocate are based on the work of Deci and Ryan (1985) and promote autonomy, competence, and relatedness. We have added an additional component, that of critical analysis. A society which develops these values is likely to be more caring, less materialistic, and less selfish. In addition, it is also much more likely to be willing to act to eliminate the dangers we are facing through uncontrolled technology.

It is important to note that, in contrast to the manipulation of society's values by large concerns, there are an increasing number of examples of technology being utilized by individuals to promote major change in the world. One such example is the "We can solve the climate crisis campaign" under the leadership of Al Gore. This campaign spread its message by making sophisticated use of technological communication channels, including blogs, to recruit millions of volunteers worldwide in an attempt to force leaders to act on the issue of climate change (Goldberg, 2008). This type of action shows clearly that technology can be used effectively to promote real values and make a better world. It also shows clearly that the responsibility lies with all of us to make this happen.

References

Allen, T. D., and Rush, M. C. (1998). Examining the effects of organizational citizenship behavior on performance judgments: A field and laboratory study. *Journal of Applied Psychology*, 83, 247–260.

Amabile, T., Constance, N. H., and Steven, J. K. (2002). Creativity under the gun. *Harvard Business Review*, 80(8), 52–61.

Amichai-Hamburger, Y. (2005). Personality and the Internet. In Y. Amichai-Hamburger (Ed.), *The Social Net: Human behavior in cyberspace* (pp. 27–55). New York: Oxford University Press.

(2007). The way to maintain your privacy on the net. *The Marker*, August 15, 5.

Amichai-Hamburger, Y., McKenna, K. Y. A., and Samuel-Azran, T. (2008). Internet E-empowerment. *Computers in Human Behavior*, 24, 1776–1789.

Bakker, A. B., Schaufeli, W. B., Sixma, H., Bosveld, W., and Van Dierendonck, D. (2000). Patient demands, lack of reciprocity, and burnout: A five-year longitudinal study among general practitioners. *Journal of Organizational Behavior*, 21, 425–441.

Belensky, M., Clinchy, G., Goldberger, N., and Tarule, J. (1989). *Women's Ways of Knowing*. New York: Basic Books.

Ben-Shahar, T. (2007). *Happier*. New York: McGraw-Hill.

Ben-Ze'ev, A. (2005). "Detattachment": The unique nature of online romantic relationships. In Y. Amichai-Hamburger (Ed.), *The Social Net: Understanding human behavior in cyberspace* (pp. 115–138). New York: Oxford University Press.

Cialdini, R. B. (1984). *Influence*. New York: Quill.

Coovert, M., and Burke, J. (2005). Personality and the Internet. In Y. Amichai-Hamburger (Ed.), *The Social Net: Human behavior in cyberspace* (pp. 219–245). New York: Oxford University Press.

Covey, S. R. (1989). *The Seven Habits of Highly Effective People*. New York: Fireside, Simon & Schuster.

Csikszentmihalyi, M. (1975). *Beyond Boredom and Anxiety*. San Francisco, CA: Jossey-Bass.

Deci, E. L., and Ryan, R. M. (1985). *Intrinsic Motivation and Self-determination in Human Behavior*. New York: Plenum.

Diener, E., and Seligman, M. E. P. (2002). Very happy people. *Psychological Science*, 13, 81–84.

(2004). Beyond money: Toward an economy of well-being. *Psychological Science in the Public Interest*, 5, 1–31.

Durkheim, É. (1893). *The Division of Labor in Society*. New York: The Free Press (reprint 1997).

Easterlin, R. A. (2003). Explaining happiness. *Proceedings of the National Academy of Sciences*, 100, 11176–11183.

Elder, L., and Paul, R. (1994). Critical thinking: Why we must transform our teaching. *Journal of Developmental Education*, 18, 34–35.

Ellul, J. (1970). *The Technological Society* (J. Wilkinson, Trans.). New York: Knopf.

Erikson, E. H. (1968). *Identity: Youth and crisis*. New York: Norton.

Fisher, A. (2005). Does your employer help you to stay healthy? *Fortune*, 12, July 12, 60.

Fried, Y., and Ferris, G. R. (1987). The validity of the job characteristics model: A review and meta-analysis. *Personnel Psychology*, 40(2), 287–322.

Goldberg, S. (2008). Gore to recruit 10m-strong green army. *Guardian International* section, April 1, 19.

Goleman, D. (1995). *Emotional Intelligence*. New York: Bantam Books.

Hackman, J. R., and Oldham, G. R. (1980). *Work Redesign*. Reading, MA: Addison-Wesley.

Hamburger, Y. A. (2000). Mathematical leadership vision. *The Journal of Psychology*, 134, 601–611.

Harrington, E. M. (1966). *The Accidental Century*. Baltimore, MD: Penguin Books.

Harrison, A., Wheeler, P., and Whitehead, C. (2004). *The Distributed Workplace*. London: Spon Press/Taylor & Francis Group.

Helliwell, J. F., and Putnam, R. D. (2004). The social context of well-being. *Philosophical Transactions of the Royal Society of London*, 359(1449), 1435–1446.

Huxley, A. L. (1932). *Brave New World*. London: Chatto & Windus.

Kasser, T. (2002). *The High Price of Materialism*. Cambridge, MA: MIT Press.

Kasser, T., Ryan, R. M., Couchman, C. E., and Sheldon, K. M. (2004). Materialistic values: Their causes and consequences. In T. Kasser and A. D. Kanner (Eds.), *Psychology and Consumer Culture: The struggle for a good life in a materialistic world* (pp. 11–28). Washington, DC: American Psychological Association.

Klein, N. (2000). *No Logo*. Toronto: Random House.

Konovsky, M. A., and Pugh, S. D. (1994). Citizenship behavior and social exchange. *Academy of Management Journal*, 37, 656–669.

Lantz, A. (1985). Strategies to increase mathematics enrollments. In S. Chipman, L. Brush, and D. Wilson (Eds.), *Women and Mathematics: Balancing the equation* (pp. 329–354). Hillsdale, NJ: Erlbaum.

Layard, R. (2005). *Happiness: Lessons from a new science*. London: Penguin.
Lyubomirsky, S., King, L. A., and Diener, E. (2005). The benefits of frequent positive affect: Does happiness lead to success? *Psychological Bulletin*, 131, 803–855.
Marks, N., and Shah, H. (2005). A well-being manifesto for a flourishing society. In F. A. Huppert, N. Baylis, and B. Keverne (Eds.), *The Science of Well-being* (pp. 503–531). New York: Oxford University Press.
Mertes, L. (1991). Thinking and writing. *Middle School Journal*, 22, 24–25.
Moneta, G. B., and Csikszentmihalyi, M. (1996). The effect of perceived challenges and skills on the quality of subjective experience. *Journal of Personality*, 64, 275–310.
Orwell, G. (1949). *Nineteen Eighty-four*. London: Secker & Warburg.
Postman, N. (1982). *The Disappearance of Childhood*. New York: Vintage.
(1985). *Amusing Ourselves to Death: Public discourse in the age of show business*. New York: Viking Penguin.
(1993). *Technopoly: The surrender of culture to technology*. New York: Vintage Books.
Potter, W. J. (2001). *Media Literacy*. London: Sage.
Putnam, R. D. (2000). *Bowling Alone: The collapse and revival of American community*. New York: Simon & Schuster.
Rogers, C. (1961). *On Becoming a Person: A psychotherapist's view of psychotherapy*. London: Constable.
Ryan, R. M., and Deci, E. L. (2000). Self-determination theory and the facilitation of intrinsic motivation, social development, and well-being. *American Psychologist*, 55, 68–78.
(2007). Active human nature: Self-determination theory and the promotion and maintenance of sport, exercise, and health. In M. S. Hagger and N. L. D. Chatzisarantis (Eds.), *Intrinsic Motivation and Self-determination in Exercise and Sport* (pp. 1–19). Leeds: Human Kinetics Europe Ltd.
Ryan, R. M., Kuhl, J., and Deci, E. L. (1997). Nature and autonomy: Organizational view of social and neurobiological aspects of self-regulation in behavior and development. *Development and Psychopathology*, 9, 701–728.
Sadker, M., and Sadker, D. (1984). *Year 3: Final report: Promoting effectiveness in classroom instruction*. Washington, DC: National Institute of Education.
Seligman, M. E. P. (2002). *Authentic Happiness*. New York: Free Press.
Sheldon, K. M., and Elliot, A. J. (1999). Goal-striving, need-satisfaction, and longitudinal well-being: The Self-Concordance Model. *Journal of Personality and Social Psychology*, 76, 482–497.
Sheldon, K. M., Elliot, A. J., Kim, Y., and Kasser, T. (2001). What is satisfying about satisfying events? Testing 10 candidates' psychological needs. *Journal of Personality and Social Psychology*, 80, 325–339.
Strenger, C. (2004). *The Designed Self*. Hillsdale, NJ: The Analytic Press.
Toffler, A. (1970). *Future Shock*. New York: Bantam Books.
Turkle, S. (1995). *Life on the Screen: Identity in the age of the Internet*. New York: Simon & Schuster.

Twenge, J. M. (2002). Birth cohort, social change, and personality: The interplay of dysphoria and individualism in the 20th century. In D. Cervone and W. Mischel (Eds.), *Advances in Personality Science* (pp. 196–218). New York: Guilford Press.

Weber, M. (1958). *The Protestant Ethic and the Spirit of Capitalism*. New York: Scribner's Press.

Yukl, G. (2002). *Leadership in Organizations*, 5th edn. Upper Saddle River, NJ: Prentice-Hall.

Index